AF837494

Value Creation through Sustainable Manufacturing

Matthew Franchetti

Behin Elahi

Somik Ghose

Industrial Press, Inc.

Industrial Press, Inc.
32 Haviland Street, Suite 3
South Norwalk, Connecticut 06854
Tel: 203-956-5593, Toll-Free: 888-528-7852
E-mail: info@industrialpress.com

Library of Congress Cataloging-in-Publication Data

Franchetti, Matthew J., author.
Value creation through sustainable manufacturing / Matthew Franchetti, Behin Elahi, and Somik Ghose.
pages cm
Includes index.
ISBN 978-0-8311-3521-8 (hardcover)
1. Production management—Environmental aspects. 2. Manufacturing processes—Environmental aspects. I. Elahi, Behin, 1983– author. II. Ghose, Somik, 1977– author. III. Title.

TS155.7.F73 2015
670—dc23 2015026565

ISBN: 978-0-8311-3521-8
ISBN ePDF: 978-0-8311-9364-5
ISBN ePUB: 978-0-8311-9365-2
ISBN eMOBI: 978-0-8311-9366-9

Copyright © 2016 by Industrial Press, Inc.
All rights reserved.

This book, or any parts thereof, with the exception of those figures in the public domain, may not be reproduced, stored in a retrieval system, or transmitted in any form without the permission of the copyright holders.

Sponsoring Editor: Taisuke Soda
Managing Editor: Laura Brengelman
Project Editor: Janine Stanley-Dunham
Interior Text Designer: Janet Romano-Murray
Cover Designers: Ryan Maxwell, Janet Romano Murray

industrialpress.com
ebooks.industrialpress.com

10 9 8 7 6 5 4 3 2 1

Table of Contents

Aknowledgments

All of the authors would like to thank the Lucas County Board of Commissioners and the Lucas County Solid Waste Management Districts for their continued support of sustainability research in Northwest Ohio. The research funding provided by these groups helped to make this book possible.

Matthew Franchetti would like to acknowledge his family for their support, as well as the many engineering students at the University of Toledo who have taken part in environmental protection research under his guidance.

Behin Elahi would like to dedicate this book to her wonderful parents, beloved husband, and great teachers for believing in her and for always being there.

Somik Ghose would like to acknowledge his family and coauthors for their support throughout the course of developing this book.

Preface

Value Creation through Sustainable Manufacturing explains processes and strategies for engineers and managers interested in implementing sustainable manufacturing systems and practices in their operations. This book provides a detailed road map, integrated approaches, and industry examples to aid readers in understanding and implementing sustainability in manufacturing. It discusses a broad range of practical tactics related to sustainable design, environmental resource management, waste minimization and pollution prevention, energy efficiency and sustainable energy sources, and green supply chain management.

Value Creation through Sustainable Manufacturing provides a relevant, practical source of technical information relating to current and potential sustainable manufacturing practices. It serves as a useful and technical guide to help organizations create more sustainable manufacturing processes throughout their entire supply chain.

Chapter 1

Introduction to Sustainable Manufacturing

"Sustainability can't be like some sort of a moral sacrifice
or political dilemma or a philanthropical cause.
It has to be a design challenge."[1]

—Bjarke Ingels

1.1. The Concept of Sustainable Systems and Sustainable Development

Sustainability in its simplest terms can be defined as meeting the needs of today without compromising resources for future generations. Many manufacturing organizations view sustainability as a legal constraint and hence define it as the conformance to environmental laws and regulations. The aim of this book is to demonstrate that if an organization focuses on sustainability, it can become a competitive advantage and lead to lower operating costs, higher quality, enhanced customer satisfaction, and increased revenue. The ultimate goals of the methods and tools presented in this book will allow organizations to:

- Increase business competitiveness through reduced operating costs and reduced waste
- Protect the environment by reducing pollution
- Increase customer satisfaction through improved environmental protection
- Accelerate the ability of an organization to adapt and thrive in a changing business environment
- Improve corporate images as the organizations excel beyond competitors

The focus of this book is to empower companies and organizations to utilize sustainability as a competitive advantage to deliver higher performance. Many organizations are able to increase profits for segments of their operations by reducing costs, raw material purchases, and other operating costs. However, many companies do not have the capability to perform comprehensive sustainability implementation, owing to time constraints and lack of knowledge in the field.

It can be easy for an organization to get wrapped up in the emotions of environmental protection and sustainability, often associated with environmental activists, or "tree huggers." The real challenge is to have these initiatives make business sense for manufacturers. In other words, how will sustainability initiatives improve the profitability or operating budgets for manufacturers? That is the focus of this book, to translate sustainability into terms that appeal to corporate executive officers (CEOs) and that will motivate them to action based on the economic benefits as well as the environmental benefits. A 2013 survey conducted on manufacturing companies in northwestern Ohio that had implemented sustainability strategies indicated an average annual cost savings of $155,000 per company. These are facts and figures that motivate upper management and CEOs. The central concept is that companies can help the environment and their economics at the same time. This book provides a detailed framework and reference material for implementing sustainable manufacturing, and it contains three key features to assist organizations in these efforts:

1. Feature 1 includes definitions, benefits, and background information regarding sustainable manufacturing.
2. Feature 2 includes discussion regarding sustainable manufacturing strategies and provides the general framework for implementation and sustained success.
3. Feature 3 provides several case studies in various industries.

1.2. Sustainability and Societal Impacts

The world and manufacturing have changed significantly over the past century. The world is now a complex global economy, and many organizations are exploring a variety of international low-cost options to create, produce, and transport manufacturing goods. This can create a very complicated environmental management issue for international manufacturing organizations, as what may be a legal waste disposal alternative in one country may be illegal in another area where the organization operates. To compound matters, societies are shifting to a convenience-oriented mind-set, world populations are increasing, and subsequently waste generation is shifting, which is creating new environmental impacts.

One of the central purposes of sustainability is to reduce or eliminate the environmental impacts of individual companies and industries. An understanding of these impacts is critical when addressing environmental issues to provide direction for reduction efforts. Some of these impacts are more important than others; thus a comprehensive understanding will allow managers and engineers to focus on more serious problems. This section provides an overview of these impacts with a focus on the effects, not the sources. These impacts are:

- **Space availability:** As the world population increases and cities grow, the available space to dispose of solid waste decreases. By minimizing waste levels, disposal space will also decrease, making land available for other uses.
- **Landfill leachate:** Leachate is liquid generated from a landfill that is created from decomposing waste or created after rainwater mixes with the chemical waste in a landfill, or the liquids present in the landfill. Once the leachate enters the environment, groundwater is at risk for mixing with it near the site, which can have very negative effects. Groundwater is the source of drinking water for over 40% of the general population and up to 90% of the population in rural areas. It was formerly assumed that this source of water was not subject to contamination from leachate, owing to its depth, but recent studies have shown that this source of water can in fact become contaminated.[2]

Landfill leachate may be virtually harmless or dangerously toxic, depending on the characteristics of the material in the landfill. Typically, landfill leachate has high concentrations of nitrogen, iron, organic carbon, manganese, chloride, and phenols. Other chemicals including pesticides, solvents, and heavy metals may also be present. Modern landfill sites require that the landfill leachate be collected and treated. Since there is no method to ensure that rainwater cannot enter the landfill site, landfill sites must now have an impermeable layer at the bottom. The landfill leachate that collects at the bottom must be monitored and treated if required. This liquid can be treated in a similar manner to sewage, and the treated water can then be safely released into the environment.

- **Global warming:** One study reports that U.S. landfills are responsible for 3.8% of the global warming damage from human sources in theUnited States. Municipal solid waste landfills are the largest source of human-related methane emissions in the United States, accounting for about 25% of these emissions in 2004[3 This gas consists of about 50% methane (CH_4), the primary component of natural gas; about 50% carbon dioxide (CO_2); and a small amount of nonmethane organic compounds. In 2003, U.S. landfills generated 131.2 teragrams of CH_4 in terms of CO_2 equivalents (where a teragram is1 million metric tons). Reducing the amounts of solid waste disposed in landfills would reduce methane generation and subsequently reduce global warming.
- **Consumption of natural resources:** A large component of solid waste minimization is recycling. Recycling reduces the consumption of virgin natural resources by utilizing perceived waste materials. For example, production of recycled paper uses 80% less water and 65% less energy and produces 95% less air pollution than virgin paper production. If all Americans recycled their newspapers just one day a week, the United States would save approximately 36 million trees a year. For every 1-kilometer (4-foot) stack of paper recycled, one tree is saved and deforestation is minimized. Recycling also reduces environmental impacts of mining. For example, by recycling aluminum, the need for the raw mineral bauxite is eliminated, which in turn eliminates the need for mining and smelting.
- **Loss of habitat:** Although it is difficult to accurately quantify habitat loss, many animal species are displaced by the creation of landfills and the effects of deforestation. By minimizing waste levels and increasing recycling, available habitats for animals should not be disrupted by the development or expansion of landfills and the effects of deforestation to acquire virgin raw materials.

1.3. Sustainability as a New Paradigm in Manufacturing

Concern for the environment and high levels of sustainability in manufacturing operations is no longer optional for most organizations; it has become an expectation and requirement for most businesses based on end customer expectations, supplier certifications, and government regulations. Additionally, sustainability can lower operating costs for most businesses, as detailed in later chapters of this book. For example, McDonald's reports that 100% of its whitefish is sustainably sourced, 90% of restaurants recycle used cooking oil, and 77% recycle cardboard.[4] These examples illustrate how sustainability and environmental protection has become mainstream in global industries.

To intensify the situation and need to focus on sustainability, social media and the rapid spread of information can lead to very negative publicity for companies that are not responsible corporate citizens and engage in actions that harm the environment. For example, at BP, the *Washington Post* reported that during the first 20 days of the Deepwater Horizon oil spill in 2010, BP spent $350 million, or $17.5 million per day.[5] This figure does not include the lost goodwill and consumer confidence in the brand.

The global marketplace has shifted into a place where sustainability is an expectation of doing business in the first place. The proliferation of ISO 14001, Leadership in Energy and Environmental Design (LEED), and other environmental certifications for manufacturing organizations is rapidly promoting the focus on the environment. Companies that do not focus on environmental protection may find themselves losing market share and subjected to increased government scrutiny and fines. In the end, a focus on sustainable operations will lead to positive outcomes for the organization and help to ensure long-term success.

1.4. Motivation for Sustainable Manufacturing and Objectives

In order to better understand the motivation for sustainable manufacturing, sustainability in terms of the triple bottom line (TBL) should be defined. The TBL concept has been utilized by many organizations to measure sustainability success, and it refers to measuring outcomes in three areas: economic, environmental, and social. Sustainability, as defined by the TBL concept, is the intersection of strong performance in economic, environmental, and social initiatives, as displayed in Figure 1.1.

The TBL concept has also been defined as promoting the 3Ps: people, profit, and planet. In practical terms, TBL accounting means expanding the traditional reporting framework to take into account ecological and social performance in addition to financial performance. The TBL phrase was

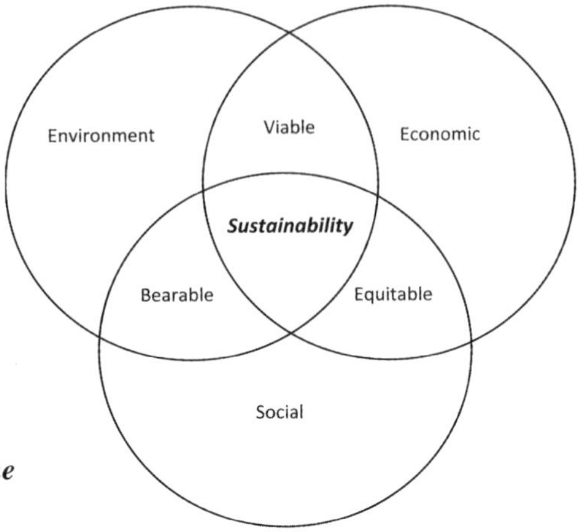

Figure 1.1. The Triple Bottom Line Intersection

coined by John Elkington in his 1998 book *Cannibals with Forks: the Triple Bottom Line of 21st Century Business.*[6] To elaborate on each measurement of the TBL, the "economic" bottom line is the financial value created by the organization for society. It differs from the traditional definition of profit in that it encompasses value to society and the community in addition to internal measures such as profit-per-share and rate of return. The "environmental" bottom line refers to long-term sustainable environmental practices. This component focuses on protecting the natural world and not harming the environment in any way in terms of waste, emissions, and destruction. It involves carefully managing consumption of energy and non-renewable resources and reducing manufacturing waste as well as rendering waste less toxic before disposing of it in a safe and legal manner. The "social" measurement of the TBL generally pertains to fair and beneficial business practices toward the workforce, the community, and the region in which a corporation conducts its business. The social component emphasizes the interrelatedness of all three components of the TBL and seeks to benefit all stakeholders (this includes employees and the community). The "social" measurement is often the most difficult to quantify because of its subjectivity, but organizations have tracked it using public surveys, outreach events, and philanthropic endeavors.

The concept of the TBL places strong emphasis on the stakeholders of an organization versus the shareholders. Stakeholders are individuals, groups, or organizations that are influenced, either directly or indirectly, by the actions of the company or organization. The TBL concept adds social and environmental values to the traditional economic measures of an organization's success. In terms of organizational decision making, the TBL concept moves beyond ranking alternatives based strictly on the financial bottom line and requires evaluating alternatives based on how they will impact the environment in terms of pollution and how the alternatives will successfully integrate into the culture of the organization or community. For example, if a company is considering implementing an office recycling program, the management team would need to evaluate each alternative program on economic factors (initial costs, program costs, revenue, payback period, and rate or return), environmental factors (waste diverted from landfills, energy usage, and changes in emissions), and social factors (capability/willingness of the employees to participate in the program and the positive and negative perceptions from the community related to the implementation of such a program). An alternative may meet the requirements of one or two of the TBL measures but must meet all three to maximize sustainability success.

In terms of sustainability and the TBL, long-term success and viability for a manufacturing organization will be found in the intersection of environmental protection, financial performance, and positive social change. Performing in the TBL intersection will lead to an organization being a good corporate citizen with happy and healthy employees and strong branding. In addition, TBL performance is becoming a requirement to doing business and will help to minimize future environmental and economic liabilities.

Based on TBL performance, the objectives of sustainable manufacturing are:

- Align sustainability efforts with corporate strategies, missions, and the customer base
- Minimize the environmental footprint of the organization (zero net harm in terms of emissions and pollution)
- Promote financial strength and success
- Promote a healthy work environment
- Maintain a low-impact life cycle of end products (production, usage, and disposal)

1.5. Regulatory Framework for Sustainable Manufacturing Practices

The governments of the world have a critical role in managing and protecting the environment, including the air, land, and water bodies. Governments are valuable institutions for resolving problems involving natural resources at both the local and global scales. This section discusses several of the key environmental laws and regulations that pertain to environmental protection in the United States.

1.5.1. Solid Waste Disposal Act of 1965[7]

The Solid Waste Disposal Act (SWDA; P.L. 89-272, 79 Stat. 992) became law on October 20, 1965. In its original form, it was a broad attempt to address the solid waste problems confronting the nation through a series of research projects, investigations, experiments, training, demonstrations, surveys, and studies. The key points of the SWDA were:

- To promote better management of solid wastes.
- To support resource recovery.
- To direct that the U.S. Public Health Service (PHS) promulgate and enforce regulations for solid waste collection, transportation, recycling, and disposal. (The U.S. Environmental Protection Agency was not formed until 1970.)
- To provide financial assistance for states to study and develop solid waste management plans.
- To provide support for research and development of improved methods of solid waste management.

The decade following its passage revealed that the SWDA was not sufficiently structured to resolve the growing mountain of waste disposal issues facing the country. As a result, significant amendments were made to the act with the passage of the Resource Conservation and Recovery Act (RCRA) of 1976.

1.5.2. National Environmental Policy Act of 1969[7]

The National Environmental Policy Act (NEPA), established in 1969 by the United States government (42 U.S.C. 4321), was one of the first laws written related to environmental protection. NEPA established a basic policy to assure that all branches of government give proper consideration to the environment prior to undertaking any major federal action that significantly affects the environment. One of the key phrases from this document that relates to recycling is "to enhance the quality of renewable resources and approach the maximum attainable recycling of depletable resources."

NEPA requires federal agencies to integrate environmental values into their decision-making processes by considering the environmental impacts of their proposed actions and reasonable alternatives to those actions. To meet this requirement, federal agencies prepare a detailed statement known as an Environmental Impact Statement (EIS). The Environmental Protection Agency (EPA) reviews and comments on EISs prepared by other federal agencies, maintains a national filing system for all EISs, and assures that its own actions comply with NEPA.

1.5.3. Resource Recovery Act of 1970[7]

Increasing concerns over protection of human health and the environment led to amendments of the 1965 SWDA, and the 1970 Resource Recovery Act (RRA) was passed. The RRA increased federal involvement with management of solid waste; it encouraged waste reduction and resource recovery and created national disposal criteria for hazardous wastes. This act was the forerunner of the Resource Conservation and Recovery Act of 1976. The two key points of the RRA were:

1. Directed that the nation would change its emphasis from solid waste disposal to recycling and energy recovery
2. Required the U.S. Public Health Service to investigate and report on the disposal of hazardous waste in the nation

This was an important guidance document for the early stages of solid and hazardous waste management. The RRA also marked the birth of the EPA, as it was formed in the interim.

1.5.4. Clean Air Act of 1970[7]

Congress established much of the basic structure of the Clean Air Act (CAA) in 1970 and made major revisions in 1977 and 1990. Dense, visible smog in many of the nation's cities and industrial centers helped to prompt passage of the 1970 legislation at the height of the national environmental movement. The subsequent revisions were designed to improve its effectiveness and to target newly recognized air pollution problems such as acid rain and damage to the stratospheric ozone layer.

To protect public health and welfare nationwide, the CAA requires the EPA to establish national ambient air quality standards for certain common and widespread pollutants based on the latest science. The EPA has set air quality standards for six common "criteria pollutants": particulate matter (also known as particle pollution), ozone, sulfur dioxide, nitrogen dioxide, carbon monoxide, and lead.

States are required to adopt enforceable plans to achieve and maintain air quality meeting the air quality standards. State plans also must control emissions that drift across state lines and harm air quality in downwind states.

1.5.5. Clean Water Act of 1972[7]

The Clean Water Act (CWA) is the primary federal law in the United States governing water pollution. It was passed in 1972 with the objective that the federal Water Pollution Control Act, commonly referred to as the Clean Water Act, restore and maintain the chemical, physical, and biological integrity of the nation's waters by preventing point and nonpoint pollution sources, providing assistance to publicly owned treatment works for the improvement of wastewater treatment, and maintaining the integrity of wetlands.

The CWA made it unlawful to discharge any pollutant from a point source into navigable waters, unless a permit was obtained. The EPA's National Pollutant Discharge Elimination System (NPDES) permit program controls discharges. Point sources are discrete conveyances such as pipes or man-made ditches. Individual homes that are connected to a municipal system, use a septic system, or do not have a surface discharge do not need an NPDES permit; however, industrial, municipal, and other facilities must obtain permits if their discharges go directly to surface waters.

1.5.6. Resource Conservation and Recovery Act of 1976[7]

The Resource Conservation and Recovery Act (RCRA) is the primary law of the United States governing the disposal of solid and hazardous waste. Congress passed the RCRA on October 21, 1976, to address the increasing problems the nation faced from our growing volume of municipal and industrial waste. The RCRA, which amended the Solid Waste Disposal Act of 1965, set national goals for:

- Protecting human health and the environment from the potential hazards of waste disposal
- Conserving energy and natural resources
- Reducing the amount of waste generated
- Ensuring that wastes are managed in an environmentally sound manner

The RCRA also included directives that the U.S. EPA establish regulations to control solid waste disposal. To achieve these goals, the RCRA established three distinct yet interrelated programs:

- The solid waste program, under RCRA Subtitle D, encourages states to develop comprehensive plans to manage nonhazardous industrial solid waste and municipal solid waste, sets criteria for municipal solid waste landfills and other solid waste disposal facilities, and prohibits the open dumping of solid waste.
- The hazardous waste program, under RCRA Subtitle C, establishes a system for controlling hazardous waste from the time it is generated until its ultimate disposal— in effect, from "cradle to grave."
- The underground storage tank (UST) program, under RCRA Subtitle I, regulates underground storage tanks containing hazardous substances and petroleum products.

The RCRA banned all open dumping of waste, encouraged source reduction and recycling, and promoted the safe disposal of municipal waste. It also mandated strict controls over the treatment, storage, and disposal of hazardous waste. The first RCRA regulations, "Hazardous Waste and Consolidated Permit Regulations," published in the Federal Register on May 19, 1980 (45 FR 33066), established the basic "cradle to grave" approach to hazardous waste management that exists today. The RCRA focuses only on active and future facilities and does not address abandoned or historical sites that are managed under the Comprehensive Environmental Response, Compensation, and Liability Act—commonly known as Superfund.

1.5.7. Hazardous and Solid Waste Amendments of 1984[7]

The RCRA was amended and strengthened by Congress in November 1984 with the passage of the federal Hazardous and Solid Waste Amendments (HSWA). These amendments to RCRA required the phasing out of land disposal of hazardous waste. Some of the other mandates of this strict law include increased enforcement authority for EPA, more stringent hazardous waste management standards, and a comprehensive underground storage tank program. Key points of the HSWA are:

- Directed the U.S. EPA to revise criteria for landfills that receive hazardous household waste or small quantities of industrial hazardous waste
- Required treatment of all contaminated surface water running off landfills
- Included methods of disposing of wastewater sewage sludge at landfills in the Clean Water Act as amended
- Increased the enforcement authority of the EPA
- Provided more stringent hazardous waste management standards
- Created a comprehensive underground storage tank program

The HSWA marked the most significant set of amendments to the RCRA with a complex law with many detailed technical requirements. Additional restrictions on land disposal and the inclusion of small-quantity hazardous waste generators (those producing between 100 kilograms [220 pounds] and 1,000 kilograms [2,200 pounds] of waste per month) in the hazardous waste regulatory scheme were added. The EPA was directed to issue regulations governing those who produce, distribute, and use fuels produced from hazardous waste, including used oil. Under the HSWA, hazardous waste facilities owned or operated by federal, state, or local government agencies must be inspected annually and privately owned facilities must be inspected at least every two years. Each federal agency was required to submit to the EPA an inventory of hazardous waste facilities it has possessed in its history. The HSWA also imposed on the EPA a timetable for issuing or denying permits for treatment, storage, and disposal facilities; required permits to be for fixed terms not exceeding 10 years; terminated in 1985 the "interim status" of land disposal facilities that existed prior to the RCRA's enactment, unless they met certain requirements; required permit applications to be accompanied by information regarding the potential for public exposure to hazardous substances in connection with the facility; and authorized the EPA to issue experimental permits for facilities demonstrating new technologies.

The EPA's enforcement powers were increased, the list of prohibited actions constituting crimes was expanded, penalties were increased, and the citizen suit provisions were expanded. Other provisions prohibited the export of hazardous waste unless the government of the receiving country formally consented to accept it; created an ombudsman's office in the EPA to deal with RCRA-associated complaints, grievances, and requests for information; and reauthorized the RCRA through FY88 at a level of about $250 million per year. Finally, the HSWA called for a National Ground Water Commission to assess and report to Congress in two years on groundwater issues and contamination from hazardous wastes. The commission was never funded and never established, however.

1.5.8. Pollution Prevention Act of 1990[7]

The Pollution Prevention Act (PPA) of 1990 was passed as part of the Omnibus Budget Reconciliation Act of 1991. The measure declared pollution prevention to be the national policy and directed the EPA to undertake a series of activities aimed at preventing the generation of pollutants, rather than controlling pollutants after they are created. Matching grants were authorized for states to establish technical assistance programs for businesses, and the EPA was directed to establish a Source Reduction Clearinghouse to disseminate information. The act also imposed new reporting requirements on industry. Firms that were required to file an annual toxic chemical release form under the Emergency Planning and Community Right-to-Know Act of 1986 must also file a report detailing their source reduction and recycling efforts over the previous year. A more complete description of the act, which addresses air and water pollution as well as waste, is provided in the first section of this report.

1.5.9. Federal Facilities Compliance Act of 1992[7]

The Federal Facilities Compliance Act (FFCA) of 1992 amended the RCRA with the primary purpose of ensuring a complete and unambiguous waiver of sovereign immunity with regard to the imposition of administrative and civil fines and penalties on federal facilities. It requires compliance with all federal, state, interstate, and local requirements in the same manner and extent as any person. The objectives of the FFCA bring all federal facilities into compliance with applicable federal and state hazardous waste laws, of waiving federal sovereign immunity under those laws, and of allowing the imposition of fines and penalties. The law also requires the U.S.

Department of Energy to submit an inventory of all its mixed waste and to develop a treatment plan for mixed waste. The FFCA has 10 sections, concerning extent of waiver of sovereign immunity, application of the RCRA to radioactive mixed wastes, application to public vessels, to waste munitions, and to federally owned treatment works. The language of the RCRA waiver of sovereign immunity prior to the FFCA was unclear on whether federal facilities had to pay civil and administrative penalties for violations of hazardous waste provisions. Cases have involved the determination of whether federal facilities have to pay penalties, even though there was no waiver of sovereign immunity. Legislatively, there were many failed attempts at passing the act. But owing to increasing support for environmental movements and frustration at state regulatory agencies that could not hammer federal facilities with fines and penalties, the act was passed in 1992. Key points of the FFCA are:

- All federal agencies are subject to all substantive and procedural requirements of federal, state, and local solid and hazardous waste laws in the same manner as any private party.
- The sovereign immunity of the United States is expressly waived in all such cases.
- "Substantive and procedural requirements" of such law include all administrative orders, civil and administrative fines and penalties, and reasonable service charges imposed for issuing and reviewing permits, plans and studies, and inspecting facilities.
- Employees, officers, and agents of the United States may not be liable for civil penalties under any such law for actions committed within the scope of that person's official duties, but such persons may be liable for criminal penalties.
- The Administrator of the EPA is authorized to commence an administrative enforcement action against any federal agency or department in the same manner as against a private party.
- Agencies must reimburse the EPA for the required annual inspections of agency hazardous waste facilities and for the EPA to conduct a comprehensive groundwater monitoring evaluation at the first inspection of each such site conducted after October 6, 1992 (unless such an evaluation has been conducted within the preceding year).

- In consultation with the Secretary of Defense, the EPA is required to propose regulations identifying when military munitions become hazardous waste and providing for the safe transportation and storage of such waste.
- Federally owned wastewater treatment works are not to be considered hazardous waste facilities if most of the water treated consists of domestic sewage, and certain other specified requirements are met. Introduction of a hazardous waste into a federally owned wastewater treatment works is prohibited.

1.5.10. Landfill Disposal Program Flexibility Act of 1996[7]

The Land Disposal Program Flexibility Act of 1996 provided regulatory flexibility for land disposal of certain wastes, as amendments to the RCRA. This act exempts hazardous waste from RCRA regulation if it is treated to a point where it no longer exhibits the characteristic that made it hazardous and is subsequently disposed in a facility regulated under the Clean Water Act or in a Class I deep injection well regulated under the Safe Drinking Water Act of 1974 (with amendments in 1986 and 1996). A second provision of the bill exempted small landfills located in arid or remote areas from groundwater monitoring requirements, provided there is no evidence of groundwater contamination.

References

1. "Bjarke Ingels," Brainy Quote, Xplore, Inc., www.brainyquote.com/quotes/authors/b/bjarke_ingels.html, retrieved May 29, 2015.

2. "Water Crisis—Water Facts," Water.org, http://water.org/water-crisis/water-facts/water/, retrieved July 29, 2014.

3. "Climate Change," U.S. Environmental Protection Agency, www.epa.gov/climate change/ retrieved July 29, 2014.

4. "2012–2013 Corporate Social Responsibility & Sustainability Report," McDonald's Corporation, www.aboutmcdonalds.com/mcd/sustainability/sustainability_CR_reports.html, retrieved August 1, 2014.

5. Mufson, S. (November 22, 2010), "Gulf Spill Reports: Federal Agencies, Oil Companies Were Unprepared for Cleanup," Washington Post.

6. Elkington, J., Cannibals with Forks: The Triple Bottom Line of 21st Century Business (North Mankato, MN: Capstone, 1998), 2.

7. "Laws and Regulations," U.S. Environmental Protection Agency, www2.epa.gov/laws-regulations/, retrieved August 7, 2014.

Chapter 2

Proactive Sustainable Manufacturing Road Map

2.1. Priorities for Industrial Implementation

Sustainability in a manufacturing environment can be a delicate balance between environmental protection, cost, and efficiency, but when an organization proactively considers sustainability while developing manufacturing processes, all three objectives can be met. The priorities for the industrial implementation of sustainability relate to the triple bottom line (TBL) and include:

1. Environmental protection
2. Economic viability
3. Social and workforce integration

The "sustainability road map" for manufacturing organizations has its roots in environmental protection. In terms of general priorities, this includes closely monitoring, controlling, and reducing hazardous chemical usage, solid waste generation, air and water emissions, raw materials usage/scrap; environmental certifications; supplier relationships; transportation and logistical impacts; landfilling; and regulatory compliance.

In terms of economic viability, the priorities focus on cost related to sustainability initiatives and costs of regulatory compliance. Oftentimes this involves detailed cost comparisons between current practices and more environmentally sustainable practices.

In terms of social and workforce integration, the focus tends to be on the workforce and customer preference. A key question related to any new business practice or policy involves the acceptance by the workforce and customer base. If the workforce supports an initiative, it will significantly enhance its long-term success and integrate the new practice into the "culture" of the organization. Additionally, many customers are now investigating the environmental practices of organizations from which they purchase goods; the customer base is growing that strongly considers an organization's environmental record and practices before making a purchase decision. This is at both a corporate level and an individual one.

The remainder of this chapter provides details on the sustainability road map, including the use of six sigma to guide process change and improvement, sustainability metrics, and an introduction to performance benchmarking to foster a culture of continuous improvement.

2.2. Six Sigma Methodology and Measuring Variations in Manufacturing Systems

Six sigma is a comprehensive and flexible system for achieving, sustaining, and maximizing business success. It is uniquely driven by a close understanding of customer needs; disciplined use of facts, data, and statistical analysis; and diligent attention to managing, improving, and reinventing business processes. Six sigma strives to improve quality, productivity, and bottom-line financial performance.[1]

2.2.1. The Six Sigma Approach to Process Improvement

The six sigma approach provides a methodology to achieve these successes and provides a system to base any improvement initiative, including a sustainability program. Six sigma relies heavily on data, facts, and the use of statistical tools to study whether an improvement has been made. These statistical tools are a powerful way to conduct experiments and compare data and to provide important information about a process to find the causes of problems and draw conclusions. Six sigma methodologies are broken into six fundamentals as related to sustainability:

1. Define products or services.
2. Know the stakeholders and customers and their critical needs.
3. Identify processes, methods, and systems to meet stakeholders' critical needs.
4. Establish a process of doing work consistently.

5. Error-proof process and eliminate waste.
6. Measure and analyze performance.

To achieve these fundamental goals, six sigma uses the five-step Define, Measure, Analyze, Improve, and Control (DMAIC) methodology described below:

Define

- Define and clarify the project goal and timeline with the focus placed on the patients.
- Establish the team.

Measure

- Define the current state and current processes, including the development of a process flow map to baseline the system and to identify any bottlenecks and establish current financial and environmental performance.
- Collect and display data, including task time, resources required, and process statistics.

Analyze

- Determine process capability and speed utilizing statistical tools and charts.
- Determine sources of variation and subsequent time bottlenecks.
- Identify and quantify value-added and nonvalue-added activities.

Improve

- Generate ideas, including
    - Methods to reduce, reuse, and recycle waste
    - Cost-effective environmental options
- Process changes.
- Conduct experiments and validate improved processes.
- Develop action plans and standard operating procedures.

Control

- Develop control plan.
- Monitor performance.
- Mistake-proof processes.

Six sigma has three overall targeted solutions: process improvement, process design/redesign, and process management. Process improvement refers to a strategy of fixing a problem while leaving the basic structure of the process intact. Process design/redesign is a strategy not to fix a process but to replace the process to fix the problem.[1]

Process management refers to integrating six sigma methods into everyday business, including:

- Processes are documented and managed "end-to-end," and responsibility has been assigned in such a way as to ensure cross-functional management of critical processes.
- Customer requirements are clearly defined and regularly updated.
- Measures of outputs, process activities, and inputs are thorough and meaningful.
- Managers and associates use the measures and process knowledge to assess performance in "real time" and take action to address problems and opportunities.
- Process improvement and process design/redesign are used to constantly raise the company's levels of performance, competitiveness, and profitability.

The remainder of this chapter discusses the general six sigma approach as it relates to sustainability for manufacturing operations from the systems perspective.

2.2.2. Road Map for Implementation

The approach presented in this book is based on the systems approach. The systems approach is a problem-solving philosophy that focuses on a holistic view of an organization by analyzing the linkages and interactions between the elements that compose the entire system. A *system* can be defined as group of interacting, interrelated, or interdependent elements forming a complex whole coordinated to achieve a stated purpose or goal. The systems approach emphasizes that the best method to understanding a problem is to understand the individual parts in relation to the whole. From a macro view, a system is composed of inputs, processes, and outputs all revolving around accomplishing a given goal or goals. The definition and clear understanding of this goal is critical to defining the system in terms of its processes, re-

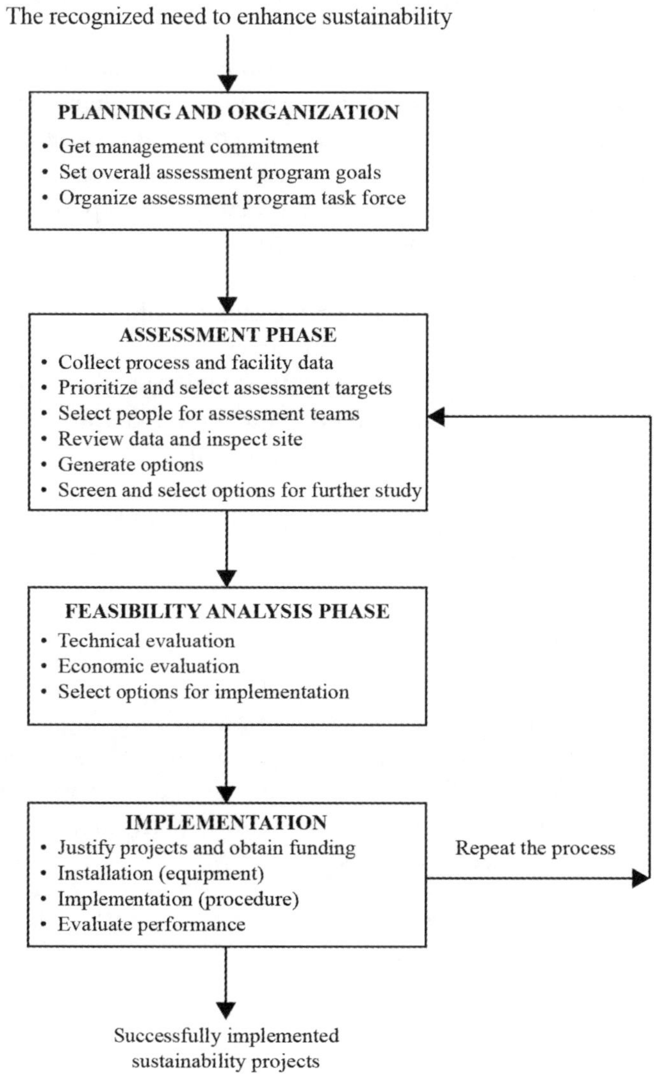

Figure 2.1. The Systems Approach to Sustainability

quired inputs, and desired outputs. For example, there will be very different systems for an organization that produces solar cells versus an organization that provides food services. The key benefit of the systems approach to sustainability is that it addresses the problem from a business standpoint, consistently focusing on the organization's goals, and confronting the problem at every stage of the supply chain.

Traditional approaches tend to address the issue of sustainability only at the end of the process, when determining how to cost effectively remove or mitigate the waste from the facility. Many organizations also manage pollution and energy usage as compartmentalized "problems" that are managed separately from their core processes. The central issue with this traditional approach is that by focusing on these individual outcomes, overall system optimization cannot be achieved. The systems approach addresses issues at all phases of the supply chain, from raw material procurement to the design of environmentally friendly processes that reduce emissions and pollution. Figure 2.1 provides an overview and road map of the system as it relates to business processes and sustainability.

The remainder of this book will explore and expand upon this sustainability road map by providing methods, strategies, tools, and examples and enhance sustainability for manufacturing organizations.

2.3. Sustainability Metrics, Indicators, and Indices for Manufacturing Industries

Accurately measuring sustainability performance and tracking trends is a critical step for a successful program in any manufacturing organization. Measurement is needed before a company can begin to manage and improve the solid waste problem. Without knowing a starting point, or baseline measurement, it is very difficult to develop a plan to meet organizational sustainability goals. This provides an overview of the various metrics to evaluate and track sustainability performance. The metrics can be refined into three categories:

- Environmental impact
- Business and financial performance
- Voice of the customers and stakeholders

The remainder of this chapter will address indices, metrics, and benchmarking with emphasis on their role in developing a strong sustainability program.

2.3.1. Sustainability Indices

Measuring and tracking the sustainability performance of a manufacturing company is critical in the effort to successfully manage and reduce pollution. For any process improvement, an accurate baseline must be created and monitored to measure success toward meeting a goal. Traditionally, sustainability efforts are reported by total output. For example, the plant generated 150 tons of solid waste this month; of this amount, 15 tons, or 10%, were recycled. This measurement approach has a very serious shortcoming. The primary flaw is that it does not consider production levels or resource inputs. For example, if two similar manufacturing plants generated the same amounts of waste but the second plant has only half the production volume, the second plant is not doing as good a job of managing its waste streams. Several other approaches have been developed to compensate for these flaws and allow for an equal apples-to-apples comparison. Five of the most common metrics by which sustainability can be measured are:

1. Absolute measures
2. Measures indexed to outputs based on production rates
3. Measures indexed to inputs based on production rates
4. Measures indexed to throughput
5. Measures indexed to a production activity

Absolute measures are not normalized based on activity; they are the actual amounts recorded, as in the 150 tons generated in a given month presented earlier. Measures indexed to production-based output provide a ratio of waste based on production volumes for a given period of time. To calculate a metric based on production rates, one would divide the waste by final output for a given period; for example, with the 136 metric tons (150 tons) generated in a given month, if an auto company produced 125,000 vehicles, its solid waste metric would be 136 metric tons/125,000 vehicles, or 0.00109 metric tons (0.0012 tons) per vehicle. This production output–based metric allows for a more accurate comparison during periods of fluctuating activity when demand may rise and fall. Measures indexed to a production-based input, measures indexed to throughput, and measures indexed to a production activity are calculated in a similar manner except they are normalized based on input (raw materials used, as in steel production per month), throughput (liters/gallons per month, as in petroleum processing), and activity (number of customer orders per month), respectively. The choice of index should be selected based on the most appropriate and easily recordable activity for the given organization. Indexed metrics are very useful and allow for apples-to-apples comparisons when activity levels change.

2.3.2. Sustainability Metrics

Common sustainability metrics relate to the TBL in terms of environmental protection, financial performance, and social well-being. From an environmental protection standpoint, this includes:

- Solid waste generation
- Recycling percentages
- Percentage of waste to landfills
- Hazardous waste generation
- Hazardous chemical usage
- Scrap generation
- Greenhouse gas (GHG) emissions
- Air emissions
- Wastewater generation
- Energy usage
- Landfill disposal methods and amounts

These environmental metrics are tracked using the most appropriate index-based measurement that fits best for the manufacturing organization.

From a financial performance point of view, rates of return, payback period, sales, market share, and revenue are typically tracked. From a business standpoint, individual sustainability projects are evaluated based on their impact to the bottom line of the organization. An understanding of the financial benefits of sustainability projects is critical in determining, evaluation, comparing, and selecting projects. In addition, a thorough understanding of the financial impact of the project will aid in promoting the project to upper management and other stakeholders. From a financial standpoint, the three areas in which sustainability projects are evaluated are:

- Initial investment
- Payback period (and discounted payback period)
- Internal rate of return

The initial investment consists of the start-up funds required to begin a given sustainability program. This includes the cost for recycling bins, recycling provider fees, recycling equipment costs (balers, grinders, or electric hand dryers found in restrooms), and training costs. The payback period is the period of time (usually given in years) required for the project's profit or other benefits to equal the project's initial investment.
The equation is:

$$Payback\ Period = \frac{Cost}{Uniform\ Annual\ Benefits}$$

The payback period measures how long the project will take to recoup the initial investment, or in other words gauges the rapid return of investment for an organization. Many organizations use the payback period as a litmus test to screen projects based on a predetermined threshold, say, three to four years for many companies. There are some limitations to the payback period, such as:

- The payback period is an approximation, not an exact economic analysis.
- All costs and all profits (or savings) of the investment before payback are included without considering differences in timing.
- All economic consequences after the payback period are ignored.
- Being an approximation, the payback period may not select the optimal project.

The internal rate of return (IRR) is the interest rate at which the present worth and equivalent annual worth of a project are equal to zero. Another way to think about the IRR is the annualized interest rate that a project earns over its life. In most cases, organizations have a predetermined minimum attractive rate of return (MARR), which is the minimum interest rate that the organization could accept as the return on a project and still remain profitable. For a given project, if the IRR is greater than or equal to the MARR, it is a profitable decision to accept the project. To solve IRR, the net present worth (NPW) is set equal to zero, as shown in the following equation:

$$NPW = \sum_{t=1}^{n} \frac{C_t}{(1+r)^t} = 0$$

Where:

t = the time of the cash flow

N = the total time of the project

r = the discount rate (the rate of return that could be earned on an investment in the financial markets with similar risk)

C_t = the net cash flow (the amount of cash) at time t

Sustainability projects should be evaluated based on these three areas: the initial investment, the payback period, and the IRR. The initial investment is important to determine and allocate the start-up funds an organization has available to begin the project. This is needed to determine starting point for the sustainability efforts. The payback period and the IRR measure the "success" of the project in financial terms.

From the standpoint of social and employee well-being, several tools exist to measure performance. Typically, surveys are used to gauge employee and customer satisfaction. Mail-based or Internet-based surveys are typically used as a low-cost option. At a higher cost, panels may be formed to "interview" groups of employees or customers to gauge their opinions toward the company. The results can be tracked over time to measure performance in this area.

2.4. Process and Performance Benchmarking

Measuring and comparing performance is a critical step to improvement. Benchmarking is the process of measuring an organization's internal processes, then identifying, understanding, and adapting outstanding practices from other organizations considered to be best-in-class.[2]

Benchmarking can be internal or external. Internal benchmarking involves an organization examining processes and instituting best practices within its own organization. External benchmarking examines the best practices of competitors or other similar organizations. Internal benchmarking is typically easier to complete, as companies have easier access to their own information and data. On the other hand, external benchmarking is more difficult and time consuming to collect, but it offers an outside perspective regarding performance. Gathering the external information may pose a challenge, but it can help determine "best in class" and to establish reach goals.

The process of benchmarking can be accomplished in three steps:[3]

1. Identify the processes in the business that are important for achieving the business goals.
2. Survey the industry and similar industries for the best performers.
3. Establish the benchmarks for each key quantity by choosing the best numbers from each data pool.

From an internal benchmarking standpoint, top performing business units or facilities should be selected for comparison. For example, General Motors has converted 109 of its manufacturing facilities to zero-landfill sites.[4] The current GM facilities that are not zero-landfill sites could use these sites to benchmark and compare their performance for improvement.

Conversely, from an external benchmarking standpoint, competitors like Ford or Chrysler could compare their facility performance to GM to identify methods and tools to convert to zero landfill. Similar benchmarking can be conducted for GHG emissions, wastewater emissions, scrap rates, and energy usage.

References

1. Pande, Peter, Newman, Robert,, and Cavanagh, Roland, introduction to *The Six Sigma Way: How GE, Motorola, and Other Top Companies Are Honing Their Performance* (New York: McGraw-Hill, 2000), p. 5.

2. "What Is Benchmarking?" Benchmarking Exchange, www.benchnet.com/wib.htm, retrieved August 21, 2014.

3."How to Benchmark Your Business," Demand Media, http://smallbusiness.chron.com/benchmark-business-23855.html, retrieved August 22, 2014.

4. "GM Achieves Zero Waste-to-Landfill in 109 Factories," Good News Network, www.goodnewsnetwork.org/gm-achieves-zero-waste-to-landfill-in-109-factories/, retrieved August 22, 2014.

Chapter 3

Management Strategies and Tactics for Sustainable Manufacturing

3.1. Systems Modeling

Systems modeling is a problem-solving philosophy that focuses on a holistic view of an organization by analyzing the linkages and interactions between the elements that compose the entire system. The systems approach is a framework based on the belief that the component parts of a system will act differently when the systems relationships are removed. Considering such an approach, component parts should also be viewed in isolation. The only way to fully understand why a problem or element occurs and persists is to understand the part in relation to the whole. The definition and a clear understanding of this goal are critical to defining the system in terms of its processes, required inputs, and desired outputs. The key benefit of the systems approach to sustainability is that it addresses the problems from a business standpoint, consistently focusing on the organization's goals and confronting the problem at every stage of the supply chain. Traditional approaches tend to address issues only at the end of the process, when determining how to cost effectively remediate environmental issues. The systems approach confronts issues at all phases of the supply chain, from procuring raw materials to designing environmentally friendly processes that reduce solid waste. To that end, the complete life cycle of the product and process is analyzed for potential environmental improvements, areas such as raw material wastes, scrap rates, material handling wastes (cardboard), and end use.

Defining the terms of a system will provide additional insights into the interactions and relationships as they relate to solid waste analysis and minimization. The starting point of any system is to clearly define its goal or purpose. Once the goal is defined, the required inputs, outputs, and process designs can be determined concurrently, focusing on the system goals. The inputs, also known as the factors of production, are land, capital, and labor. This chapter examines and explains four system models that relate to sustainability for manufacturing organizations. Each model is explained in more detail in the following sections. The four models are:

1. Cost-benefit models
2. Optimization models
3. Simulation models
4. Forecasting models

3.1.1. Cost-Benefit Models

Cost-benefit modeling is an analytical method of examining projects for investment purposes. The method has three steps:[1]

1. Identify the financial value of expected project cost and benefit variables.
2. Analyze the relationship between expected costs and benefits using simple or sophisticated selection techniques.
3. Make the investment decision.

Cost-benefit analysis typically consists of determining payback periods and internal rates of return, as discussed previously in chapter 2 (section 2.3.2., "Sustainability Metrics"). For many organizations, cost-benefit models and analyses are key aspects of the decision process to accept new projects. New projects, sustainability related or otherwise, must be justified economically before being considered for implementation. This may range from installing a new cardboard baler to mitigating greenhouse gas (GHG) emissions or swapping cutting fluids to a less hazardous version.

3.1.2. Optimization Models

Optimization modeling is an area of applied mathematics that is concerned with maximizing or minimizing a certain function under constraints. The field uses linear or nonlinear programing and other tools to mathematically define a problem, quantify constraints, and then solve for the best possible solution to maximize the value for an organization. For example, a dynamic nonlinear programming optimization model of water use in North America's Rio Grande basin was explored in 2008; the model optimized the net present value of the basin's total economic benefits subject to constraints on equity, sustainability, hydrology, and institutions.[2] It was applied to assess impacts of a two-tiered pricing program that complies with recently implemented drinking water quality standards for the basin's two largest U.S. cities (Albuquerque, New Mexico, and El Paso, Texas), and the findings provided a framework for designing water-pricing programs that comply with environmental regulations.[2]

3.1.3. Simulation Models

Simulation models involve creating an electronic computer-based version of a physical system to better understand process times and bottlenecks and to conduct "what if" analyses before making any changes to the physical system. Simulation modeling allows engineers and managers to quickly make changes to a model to measure positive and negative changes in the overall system without manipulating the physical system. Simulation models provide additional information to decision makers before making actual changes to the physical system, which are typically more time and cost intensive. For example, in Canada researchers used simulation modeling to measure changes to the maintenance of the economic, biological, and physical components of an agriculture system as related to farming.[3] The purpose of the model was to determine the most economical and environmental management options for farming operations in the region.[3]

3.1.4. Forecasting Models

Forecasting models are mathematical methods used to estimate and predict future events based on historical information and data. These models are used in a wide variety of fields of economics, business, politics, agriculture, weather, and environmental protection. There are a number of forecasting methods; the most common methods are:

- Moving average
- Weighted moving average
- Exponential smoothing
- Autoregressive moving average
- Extrapolation
- Linear prediction and regression modeling

Many companies and countries use various forecasting methods to help predict future GHG emissions. In 2013, Canada forecasted the country's GHG emissions and economic growth to strike a balance between environmental protection and job growth.[4] The study applied the moving average forecasting technique and found that Canada's GHG emissions decreased by 4.8% and the economy grew by 8.4% from 2005 to 2011. Canada then used this information to predict a future per capita GHG (total greenhouse gas emissions divided by the population of Canada) decrease from 22.9 tonnes in 2005 to 20.0 tonnes per person in 2020.[4]

3.2. Systems Assessment Tools

Sustainable manufacturing strategies and tactics demonstrated by companies are essential factors to increasing business performance and competitiveness. Sustainable manufacturing strategies offer insight into the overall approach taken by companies, but they can lack practical support for implementation. Considering this issue, examining the gap between strategic direction and practices to extract the mechanisms behind the practices and formulate sustainable manufacturing tactics (which provide information on how specific improvements can be implemented) is very critical. Applying proper system assessment tools can guide manufacturers to undertake the sustainability journey in a structured, systematic, and cross-functional way through the steps of factory modeling, resource flow (material, energy, and waste) modeling, resource flow analysis, and identification of opportunities for improvement. Systems assessment tools provide guidelines for manufacturers by utilizing various sustainable manufacturing tactics, as follows:

- **Prevention (avoid usage):** Remove the unnecessary resource usage, remove unnecessary technology, align resource input profile with production schedule, switch off/standby mode when not in use, and so forth.
- **Waste reduction:** Waste collection, sorting, recovery and treatment.
- **Resource use reduction:** Optimize production schedule to improve efficiency; optimize resource input profile to improve efficiency; change set points/running load; monitor performance; control performance; change resource flow layout; change technology layout; and so forth.
- **Reuse (waste as a resource):** Synchronize waste generation and resource demand to allow reuse; reuse waste output as resource input; and so forth.

- **Substitution (new resource or technology):** Replace resource input for better one; replace technology for better one; add a highly efficient resource; add a high efficiency technology; change the way the function is accomplished; etc.

Proper systems assessment tools target critical elements of sustainable manufacturing, such as a production system, the buildings and facilities that are servicing manufacturing operations and provide heating, ventilation, air-conditioning (HVAC), lighting, power, water, and waste removal. Some parts of related process data and their sources that are recommended to be considered in such modeling and assessments are as follows:[5]

- ♦ **Building modeling/assessment: drawing the infrastructure:**
  - Building geometry/thermal zones (Sources: Factory layout and Technical drawings)
  - Construction data (Sources: Building construction materials)
  - HVAC systems (Sources: Building service system documentation)
- ♦ **Qualitative process modeling/assessment: mapping manufacturing operations and facilities:**
  - Technology (process/equipment) geometry (Sources: Pictures and manual of equipment/processes)
  - Technology layout (Sources: Factory layout and Technical drawings)
  - Technology attributes/characteristics (Sources: Process/equipment specifications)
  - Resource layout (Sources: Energy and material path/network layout)
  - Resource characteristics (Sources: Energy and material characteristics)
  - List of processes and qualitative product flow (Sources: Manufacturing routings)
- ♦ **Quantitative process modeling/assessment: modeling/assessment of manufacturing operations and facilities:**
  - Production profile (factory-wide), equipment/process operations profile, and product profile (quantitative product flow) (Sources: Production schedules)
  - Technology set point/demand profiles (Sources: Equipment and process set points, demand, and running load)

- Technology control profiles (Sources: Controls such as controllers, and valves)
- Resource usage profiles (Sources: Facility equipment and manufacturing process controls (metered data)
- Resource supply profiles (Sources: Facility equipment generation [metered data])
- Waste profiles (Sources: Facility equipment and Manufacturing process waste generation)
- Total inputs to the system (check model completeness) (Sources: Total inputs to the system such as energy/water bills and Bill of Materials (BOM))
- Energy and mass balance (for missing data) (Sources: Thermodynamics for resource transformation process)
- Link technology to HVAC system (Sources: Thermal transfer of space/building)
- Link technology to bins (waste profile, energy and mass balance) (Sources: Waste data if available)

♦ **Optimized process modeling/assessment: improvements implementation:**

- Controller functions (for simulation purpose) (Sources: Control strategy)
- Bins/recycling repositories (Sources: Recover, sort, collect, reuse, recycle)
- Modification of technology (process/equipment) (Sources: Equipment/process management or change)
- Modification to resource flow (Sources: Resource management or change)

Statistical energy consumption data for 25 industrial divisions in the United Kingdom highlight that for a number of manufacturing industries (e.g., manufacture of motor vehicles, electrical machinery, radios, and medical equipment), building associated energy (i.e., space heating and lighting) contributes to roughly 40%–60% of the total energy consumed.[6] Therefore, there is a noteworthy possibility for resource efficiency enhancement by applying suitable systems assessment tools and looking at the factory as an ecosystem, which is shown by Figure 3.1.[7]

Another piece of statistical data, which highlight the critical need of transition to sustainable manufacturing and industries by using appropriate systems assessment tools, is related to U.S. greenhouse gas emissions al-

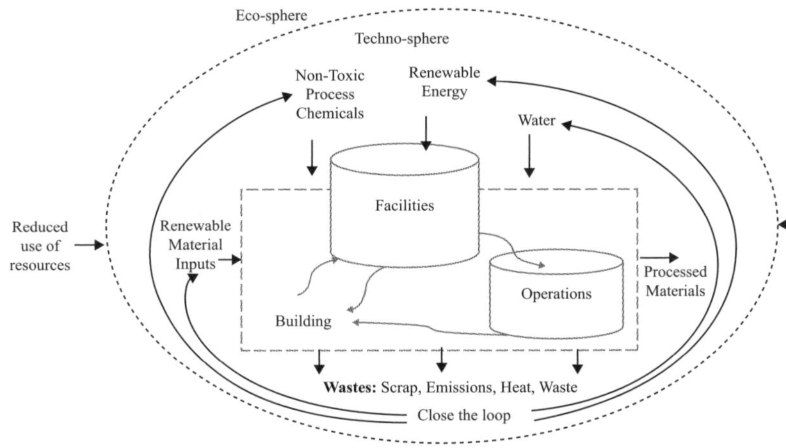

Figure 3.1. Manufacturing Ecosystem Model with the Subsystems and Resource Flows

Source: Ball, P. D., et al., "Modelling Energy Flows Across Buildings, Facilities and Manufacturing Operations."

lotted to economic sectors. Throughout the Inventory of U.S. Greenhouse Gas Emissions and Sinks report, emission approximations are categorized in different economic sectors: Residential, Commercial, Industry, Transportation, Electricity Generation, Agriculture, and U.S. Territories. Table 3.1 reviews emissions from each of these divisions based on the U.S. Environmental Protection Agency (EPA) report. Figure 3.2 also demonstrates the related trend in emissions by sector from 1990 to 2012. By means of this classification and considering GHG emissions shown in Table 3.1, emissions from electricity generation accounted for the main portion (2,064.0/6,525.6 = 31.6 percent) of U.S. greenhouse gas emissions in 2012. Transportation activities, in aggregate, accounted for the second-largest portion (1,837/6,525.6 = 28.1 percent), and emissions from industry accounted for the third-largest percentage (1,278.4/6,525.6 = 19.6 percent) of U.S. greenhouse gas emissions in 2012. The necessity of creating sustainable manufacturing activities is completely highlighted by looking at these statistical data. Moreover, the remaining 20.7 percent of U.S. greenhouse gas emissions out of 100 percent were contributed by, in sequence of significance, the agriculture, commercial, and residential sectors, in addition to emissions from U.S. Territories.[8]

Table 3.1. U.S. Greenhouse Gas Emissions Allocated to Economic Sectors (in Tg, or 1 million metric tonne of CO_2 equivalent)

Implied Sectors	**1990**	**2005**	**2008**	**2009**	**2010**	**2011**	**2012**
Electric power industry	1,866.1	2,445.7	2,401.8	2,187.0	2,302.5	2,200.9	2,064.0
Transportation	1,553.2	2,017.2	1,935.2	1,862.4	1,867.4	1,852.1	1,837.0
Industry	1,531.5	1,407.5	1,371.5	1,220.5	1,300.5	1,297.5	1,278.4
Agriculture	518.1	583.6	615.3	605.3	600.9	612.7	614.1
Commercial	385.3	370.4	379.2	381.9	376.6	378.3	352.7
Residential	345.4	371.3	365.4	357.9	360.0	353.6	321.4
U.S. Territories	33.7	58.2	49.8	47.9	58.0	57.9	57.9
Total Emissions	**6,233.2**	**7,253.8**	**7,118.1**	**6,662.9**	**6,874.7**	**6,753.0**	**6,525.6**
Land use, land-use change, and forestry (sinks)	(831.1)	(1,030.7)	(981.0)	(961.6)	(968.0)	(980.3)	(979.3)
Net Emissions (sources and sinks)	**5,402.1**	**6,223.1**	**6,137.1**	**5,701.2**	**5,906.7**	**5,772.7**	**5,546.3**

Before selecting proper systems assessment tools by manufacturing systems, it is necessary to consider three major factors. The first factor refers to the temporal characteristics, that is, if the tool assesses past development (ex-post, or descriptive) or if it is applied for forecasting future outcomes (ex-ante, or change-oriented), such as a strategy change or an improvement in a production process. The second factor is related to the focusing areas (coverage areas), for instance, if their emphasis is on the product level, or on a proposed change in policy. The third major factor signifies the assimilation of nature–society systems, that is, to what extent the tool merges environmental, social, and/or economic features. In addition to considering these factors, when a manufacturing system decides on evolution to sustainability, targets and objectives must be measured. This has posed critical challenges to the scientific and industrial community in providing efficient but reliable assessment tools.

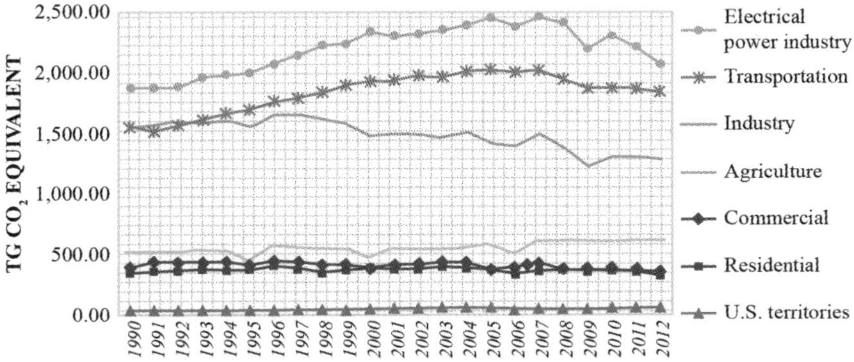

Figure 3.2. U.S. Greenhouse Gas Emissions Allocated to Economic Sectors (Tg of carbon dioxide equivalent)

**Note: Tg (teragram) CO equivalent is equal to 1 million metric tonne of carbon equivalent (MMTCE).*

As a reply to these challenges, sustainability assessment has become a speedily growing area. The amount of tools that claim they can be used for assessing sustainability has increased; concurrently many of the tools have industrialized, offering better application procedures, data, and case study experiences. Sustainability assessment has progressively become related to the group of impact assessment tools consisting of, for example, environmental impact assessment and strategic environmental assessment.[9, 10] Sustainability assessment is defined as a tool that assists decision makers to determine which actions they should or should not take in an effort to make their system more sustainable.[11] The unique characteristic of sustainability assessment when compared with other, more established procedures of impact assessment and planning is that the complex and ambiguous concept of sustainability lies at its core. The concept of sustainability assessment of a manufacturing system covers different areas containing a) metrics, indicators, and indexes, b) product/system-related assessment tools with the focus on the material, resource, waste, and energy flows of a product or service from a life cycle perspective in considering manufacturing system, and c) integrated assessment, which is a collection of tools usually focused on policy change or project implementation. In order to achieve a successful transition to sustainability, a proper systems assessment tool should be utilized and the following stages are required to be taken into consideration thoroughly:

- Developing indicators, indices, metrics, and a maturity model (through the manufacturing systems/product life cycle thinking) for sustainable manufacturing.

- Designing for sustainable products, services, and manufacturing systems: integrating environmental aspects into product design and development; performing product life cycle management and life cycle analysis/assessment, using advanced manufacturing technologies such as nanomanufacturing by collecting more information regarding material science, applying various procedures and methods for energy efficiencies, recycling, providing the system with proper information infrastructures containing advanced models and semantics for product and process, and manufacturing simulation.
- Establishing tools and standards using industry best practices for sustainable manufacturing system.
- Developing appropriate information and communication technologies (ICTs) for sustainable manufacturing: ICT for design, manufacturing, and supply chain optimization for sustainable manufacturing; large scale data modeling and semantic technologies for sustainable manufacturing.

3.2.1. Developing Indicators, Indices, Metrics

As it was mentioned earlier, manufacturers need to measure the degree of sustainability of their manufacturing processes/systems and develop indicators that are simple to implement and interpret. Since the sustainability is measured across several dimensions, the indicators used to assess the different dimensions vary and have different measuring units. In terms of choosing proper metrics, the following features should be considered by the manufacturers in their action plans:

- **Measurability:** Ensure that it is simple to compare the metrics with standards, and metrics can be simply measured using quantitative or qualitative data.
- **Ease of access to data:** Ensure that measures are based on data that are readily available in the facility.
- **Normalization:** Metrics must be normalized in regard to production.
- **Relevancy:** Metrics are directly associated with the dimensions of sustainability.

The International Organization for Standardization (ISO) 14045 takes a stance in choosing environmental indicators. It defines the principles, requirements, and guidelines for eco-efficiency assessment of production systems. In actual act, eco-efficiency is accomplished through the pursuit of three central objectives: enhancing product or service value, optimizing the use of resources in manufacturing systems, and reducing environmental impact. The Organisation for Economic Co-operation and Development (OECD) has also developed a sustainable toolkit. The OECD toolkit provides a total of 18 measurable indicators categorized according to input, operations, and output. Indicators are mainly environmentally oriented.[12] Three examples of sustainability indexes that are used by manufacturing systems are as follows:

1. Higg Index is an apparel and footwear industry self-assessment standard for evaluating environmental and social sustainability throughout the supply chain. Introduced in 2012, it was developed by the Sustainable Apparel Coalition, a nonprofit corporation established by a cluster of fashion companies and the U.S. Environmental Protection Agency (EPA).[13]

2. The Dow Jones Sustainability Indexes introduced in 1999 are a group of indexes assessing the sustainability performance of the largest 2,500 companies recorded in the Dow Jones Global Total Stock Market Index. They are the longest-running global sustainability benchmarks world wide and have become the central reference point in sustainability financing and investing for investors and enterprises alike. In 2012, S&P Indices and Dow Jones Indexes united to create S&P Dow Jones Indices.[14]

3. Standard & Poor's (S&P) and TMX Group Inc., operator of the Toronto Stock Exchange (TSX), announced the launch of a new green investing index to track clean tech companies in 2010. The S&P / TSX Clean Technology Index measures the performance of companies and manufacturing systemswhose primary business is in the development and deployment of green technologies.[15]

Here various simple sustainability indexes as samples are presented, and a method for aggregating different indicators into a sustainability index (SI) is presented. In this method each indicator is normalized and expressed in a dimensionless quantity. As it is shown in Table 3.2, the SI is measured in terms of its three main components: environmental, economic, and social components. Environmental indicators are identified by I_{1ij}, where $i = 1$ refers to element of energy, $i = 2$ refers to element of natural resources, $i = 3$ refers to materials, and $i = 4$ refers to element of emissions.

Notations are as follows:

While $i = 1$ and $j = 1$: j refers to components of the renewable energy ratio,

While $i = 1$ and $j = 2$: j refers to components of the energy intensity,

While $i = 2$ and $j = 1$: j refers to components of the water consumption,

While $i = 2$ and $j = 2$: j refers to components of land usage,

While $i = 2$ and $j = 3$: j refers to components of the water waste ratio,

While $i = 3$ and $j = 1$: j refers to components of the reused material ratio,

While $i = 3$ and $j = 2$: j refers to components of the recyclable material ratio,

While $i = 3$ and $j = 3$: j refers to components of the hazardous material ratio,

While $i = 3$ and $j = 4$: j refers to components of the wasted material ratio,

While $i = 4$ and $j = 1$: j refers to components of the direct emissions,

While $i = 4$ and $j = 2$: j refers to the components of the indirect emissions.

In addition, I_{kl} refers to economic indicators when $k = 2$; and it denotes societal indicators when $k = 3$. Other notations are as follows:

While $k = 2$ and $l = 1$: l refers to element of added value,

While $k = 2$ and $l = 2$: l refers to element of defective product ratio,

While $k = 2$ and $l = 3$: l refers to element of production diversification,

While $k = 3$ and $l = 1$: l refers to an element of the manpower turnover rate,

While $k = 3$ and $l = 2$: l refers to element of training budget ratio,

While $k = 3$ and $l = 3$: l refers to element of overtime ratio,

While $k = 3$ and $l = 4$: l refers to element of overtime intensity.

3.2.1.1. Environmental Dimension Indicators

The renewable energy ratio indicator measures the proportion of renewable energy used to the total amount of energy required. The increase in this indicator achieves more sustainability (Eq.1).

Table 3.2. List of Indicators That Are Used in Calculating an Aggregated Sustainability Index

Dimension		Indicators				ID	Optimality Direction
		Description					
1	Environmental	1	Energy	1	Renewable energy ratio	I_{111}	Max
				2	Energy intensity	I_{112}	Min
		2	Natural resources	1	Water consumption	I_{121}	Min
				2	Land usage	I_{122}	Min
				3	Wastewater ratio	I_{123}	Min
		3	Materials	1	Reduced material ratio	I_{131}	Max
				2	Recyclable material ratio	I_{132}	Max
				3	Hazardous material ratio	I_{133}	Min
				4	Wasted material ratio	I_{134}	Min
		4	Emissions	1	Direct emissions	I_{141}	Min
				2	Indirect emissions	I_{142}	Min
2	Economic	1	Added value			I_{21}	Max
		2	Defective product ratio			I_{22}	Min
		3	Production diversification			I_{23}	Max
3	Societal	1	Manpower turnover ratio			I_{31}	Min
		2	Training budget ratio			I_{32}	Max
		3	Overtime ratio			I_{33}	Min
		4	Labor intensity			I_{34}	Max

$$I_{111} = \frac{\textit{quantity of renewable energy}}{\textit{total energy consumption}} \tag{1}$$

To assess the degree of energy consumed relative to all the other resource input to the system, the following ratio is applied (Eq.2). The more energy is consumed, the less sustainable the manufacturing operation. Thus the ratio is subtracted from 1, so that the minimization of the ratio is assured to achieve an increased level of sustainability.

$$I_{112} = 1 - \frac{\$\textit{value of energy consumption}}{\$\textit{value of total inputs to production}} \tag{2}$$

Equation 3 relies on measuring the amount of water used per product. Since the amount of water may exceed the product volume, zero is used whenever the ratio turns negative.

$$I_{121} = \max\left[1 - \frac{\textit{annual water consumption}}{\textit{annual quantity produced}}, 0\right] \tag{3}$$

Calculations related to preserving the land and water resources are shown in Equation 4 and Equation 5 successively.

$$I_{122} = 1 - \frac{area\ of\ covered\ land}{total\ area\ used} \tag{4}$$

$$I_{123} = 1 - \frac{wasted\ water\ annually}{total\ water\ consumption} \tag{5}$$

Equation 6, Equation 7, Equation 8, and Equation 9 are related to reused, recyclable, hazardous, and wasted materials.

$$I_{131} = \frac{quantity\ of\ reused\ material}{total\ input\ material} \tag{6}$$

$$I_{132} = \frac{quantity\ of\ recyclable\ material}{total\ input\ material} \tag{7}$$

$$I_{133} = 1 - \frac{weight\ of\ hazardous\ material}{weight\ of\ total\ input\ material} \tag{8}$$

$$I_{134} = 1 - \frac{weight\ of\ wasted\ material}{weight\ of\ total\ input\ material} \tag{9}$$

Greenhouse gases, which comprise carbon dioxide (CO_2), methane (CH_4), hydro fluorocarbons (HFCs), nitrous oxide (N_2O), per fluorocarbons (PFCs), and sulfur hexafluoride (SF_6), may result directly from the manufacturing operations on site, or indirectly through either the process of electricity generation or the transportation activities. The two sources of emissions are formulated by Equation 10 and Equation 11, respectively. The indicator described in Equation 10 measures the weight of the gases generated and can be expressed in kilograms or other units such as pounds, teragram, and so forth of carbon dioxide equivalent (CO_{2e}), per total weight of material used. It presents the direct emissions from the manufacturing operations. In this equation, GHG refers to greenhouse gas. The ratio of indirect emissions to the total weight of finished products is expressed by Equation 11. C_{p1} and C_{p2} present the amount of emissions resulting from the generation of 1 megawatt (MW) of electricity and the transportation per unit distance, respectively. In this equation, MWh refers to megawatt-hour.

$$I_{141} = 1 - \frac{weight\ of\ \ GHG\ locally\ produced}{weight\ of\ total\ input\ material} \quad (10)$$

$$I_{142} = 1 - \frac{\left(C_{p1} \times MWh\ \ consumed + C_{p2} \times \ distance\ of\ transportation\right)}{weight\ \ of\ \ annual\ quantity\ produced} \quad (11)$$

3.2.1.2. Economic Dimension Indicators

The added value is taken as a measure of economic benefit. It is compared with the total monetary value of the total production, as formulated in Equation 12.

$$I_{21} = \frac{added\ value}{\$value\ \ of\ \ production} \quad (12)$$

Defective products indicate the inefficiency of the manufacturing system. The generation of defective products means the unconscious consumption of resources that cannot be recovered, in addition to customer dissatisfaction and lost opportunities of profit. To assess the amount of defective products, the loss function can be defined and applied (Eq.13).

$$I_{22} = 1 - loss\ function \quad (13)$$

A sustainable manufacturing system uses the same resources to generate a multitude of products, thus satisfying customer needs with a minimum level of resources. Product diversification is used in Equation 14 to assess this aspect, where N is the number of different products produced, Q is the maximum possible number of diversified products in the considered industry, and P_u is the ratio of the quantity produced of the uth product to the total production. The calculation is based on the entropy principle.

$$I_{23} = \frac{\sum_{u=1}^{N} p_u \ \times \ ln(p_u)}{ln\left(\frac{1}{Q}\right)} \quad (14)$$

3.2.1.3. Societal Dimension Indicators

A sustainable manufacturing system should provide stable job opportunities for society members. This concept is measured by Equation 15 through the number of fired or resigned employers. Firing and resignation may reflect the stability of the system and the employer-employee relationship. In addition, the ratio of training budget is shown by Equation 16. The degree of company involvement in the development of its workforce is assessed by this formula. Besides, a sustainable manufacturing system should preserve the energy and well-being of its workforce and minimize the overtime (Eq.17). Labor intensity is measured by Equation 18 through the comparison of the expenditure assigned for the labor (wages, incentive, insurance, etc.) with the total amount of expenditure by the company, including manufacturing and overhead costs.

$$I_{31} = 1 - \frac{\textit{annual layoff and resignations}}{\textit{total manpower}} \tag{15}$$

$$I_{32} = \frac{\textit{training budget}}{\textit{total expenditures}} \tag{16}$$

$$I_{33} = 1 - \frac{\textit{annual overtime}}{\textit{annual regular time}} \tag{17}$$

$$I_{34} = \frac{\textit{labor expenditures}}{\textit{total input expenditures}} \tag{18}$$

3.2.1.4. Sustainability Index

The sustainability index (SI) is calculated by two steps. First, the indicators within each of the three dimensions are algebraically added using their respective weight (W) to obtain a single measure for each. The three resulting measures are the environmental indicator (I_{EN}), the economic indicator (I_{EC}), and the societal indicator (I_{SC}). These measures can be calculated by using Equation 19, Equation 20, and Equation 21, respectively. To determine the relative importance and weight of the different criteria and their subcriteria, a proper multi-criteria decision making (MCDM) method such as analytical hierarchy process (AHP), analytic network process (ANP),

Table 3.3. Examples of Process Metrics for Sustainable Machining

Environmental impact	Energy consumption	Cost
▪ GHG emission from energy consumption of the line (tonne, teragram, lb, etc. CO_2 eq.) ▪ Ratio of renewable energy used (%) ▪ Total water consumption (tonne, lb, etc.) ▪ Mass of restricted disposals (kg, lb, etc.) ▪ Noise level outside the factory (dB)	▪ In-line energy consumption (kWh, MWh, etc.) ▪ Energy consumption on maintaining facility environment (kWh, MWh, etc.) ▪ Energy consumption for transportation ▪ Into/out of the line (kWh, MWh, etc.) ▪ Ratio of use of renewable energy (%)	▪ Labor cost ($) ▪ Cost for use of energy ($) ▪ Cost of consumables ($) ▪ Maintenance cost ($) ▪ Cost of by-product treatment ($) ▪ Indirect labor cost ($)
Operator safety	**Personal health**	**Waste management**
▪ Exposure to corrosive/toxic chemicals (incidents/person) ▪ Exposure to high energy components (incidents/person) ▪ Injury rate (injuries)	▪ Chemical contamination of working Environment (mg/m^3, lb/ft^3, etc.) ▪ Mist/dust level (mg/m^3, lb/ft^3, etc.) ▪ Noise level inside factory (dB) ▪ Physical load index (dimensionless) ▪ Health-related absenteeism rate (%)	▪ Mass of disposed consumables (kg, lb, etc.) ▪ Consumables reuse ratio (%) ▪ Mass of mist generation (kg, lb, etc.) ▪ Mass of disposed chips and scraps (kg, lb, etc.) ▪ Ratio of recycled chips and scraps

fuzzy-AHP, fuzzy-ANP, VIKOR, fuzzy VIKOR, PROMETHEE, ELECTRE, and methods for group decision making can be applied.[16]

Second, the three resulting economic, environmental, and societal indicators are taken into consideration as the three components of sustainability vector in a three-dimensional space to arrive at the SI. It is shown in Equation 22.

$$I_{EN} = \sum_{i=1}^{4}\sum_{j=1}^{n_i} W_{1ij} \times I_{1ij} \tag{19}$$

$$I_{EC} = \sum_{l=1}^{3} W_{2l} \times I_{2l} \tag{20}$$

$$I_{SC} = \sum_{l=1}^{4} W_{3l} \times I_{3l} \tag{21}$$

$$SI = \sqrt{I_{EN}^{\ 2} + I_{EC}^{\ 2} + I_{SC}^{\ 2}} \tag{22}$$

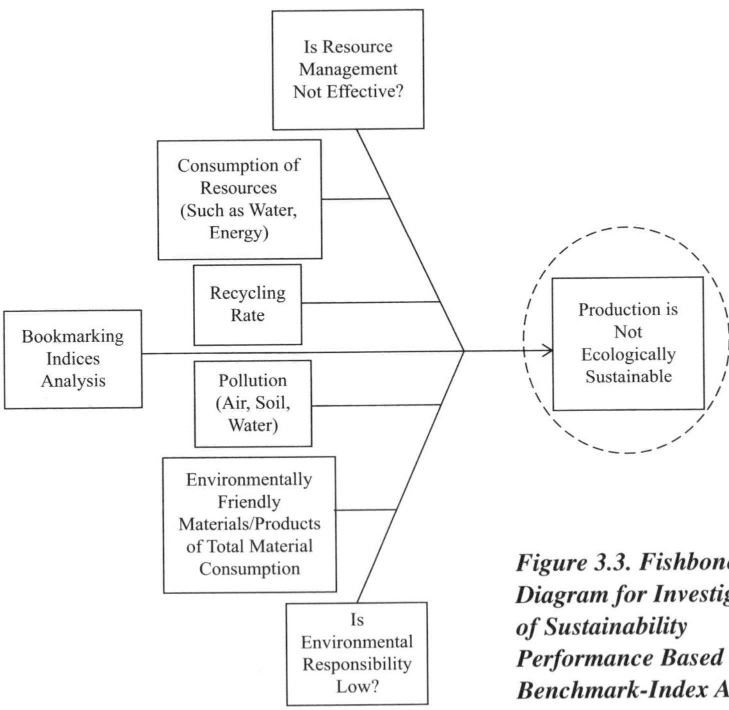

Figure 3.3. Fishbone Diagram for Investigation of Sustainability Performance Based on Benchmark-Index Analysis

Moreover, examples of process metrics for sustainable machining are shown in Table 3.3.[17] Also, some efficient sustainability indexes and sustainability best practices can be achieved by benchmarking. Benchmarking signifies a continuous process in which products and services and, in particular, processes and methods of operational functions are compared with those of several other companies or organizations. It allows the acquisition of ideas from outside the own organizational structures/manufacturing system and derivation of measures, which can be adapted and implemented as a company-specific solution. With a view of the bigger picture, a manufacturer is able to achieve a top position among all potential competitors by setting new standards in the world of the world's best. After conducting a benchmark-index analysis, the Fishbone diagram can be applied to examine the causes of failure of manufacturing systems in terms of sustainability performance (Fig. 3.3). As an example, this method has been applied by the Fraunhofer Institute for Production Systems and Design Technology in Berlin. The company focused on the integration of sustainability perspective to the underlying four perspectives of the balanced scorecard (BSC), so that benchmarking outcomes not only led to enhancing economical performances but also resulted in taking environmental requirements such as energy efficiency and social responsibility into consideration. The company could also recognize the related strengths and weaknesses by using a benchmark-index analysis.[18]

The most significant sustainability assessment tools can be categorized as management information system, information technology, and decision support system; scenario development; material flow analysis and energy analysis; life cycle assessment; risk assessment; and environmental impact assessment. These tools are introduced in the sections that follow.

3.2.2. Management Information System, Information Technology, and Decision Support System

Management information system (MIS) is a broadly practical term for a three-asset system required for an efficient system/organization management. The assets are people, information, and technology. It is worthy to note that an MIS should not only specify the status of a business's conditions but also designate why the circumstances are improving or deteriorating. Examples of the broad scope of MIS are enterprise resource planning, customer relationship management, and executive information system.[19]

Information technology (IT) is a business segment that focuses on processing and computing, containing hardware, software, telecommunications, and in general tools involved in the transmittal of information or the systems that accelerate communication.[20]

The term *decision support system* (DSS) denotes a methodology that merges decision makers' discernments with the computer's information processing capacities for enhancing the quality of decision making. A DSS is a computer-based application that characteristically includes mathematical models along with informational databases and a user interface to offer advocated decisions to manager and users. It gathers, systematizes, and evaluates enterprise data to facilitate quality enterprise decision making for management, operations, and planning. A well-designed DSS helps decision makers in compiling a different assortment of data from various sources: raw data, documents, personal knowledge from employees, management, executives, and business models. A DSS varies from an information system (IS) or an MIS, as it not only provides the user with information, databases, or reports, as does an IS or MIS, but also offers answers to user queries, that is, decisions, through its modeling module. Basically, DSS assessments help enterprises to identify and solve problems and make decisions more efficiently.

A decision-making assessment was performed by the Carnegie Institute of Technology in the late 1950s and early 1960s. The Massachusetts Institute of Technology applied computer technology to a decision-making theory in the 1960s. By the 1980s, intensive research on DSS was proceeding, and novel theories and perceptions came into sight from single-user models of DSS, containing organizational decision support system, group decision support system, and executive information system. By 1990, DSS was expanded to incorporate data warehousing and online analytical processing.[19]

3.2.2.1. Group Decision Support System, Business/System Continuity Plan, and Recovery Time Objective

Group decision support system (GDSS) technology advocates project cooperation by means of the improvement of digital communication using diverse tools and resources. These forms of programs are applied to support customized projects calling for group work, input to a group, and numerous sorts of meeting protocols. GDSS technologies can enhance the promotion of participation, assist to streamline group communications, and promote learning. GDSS is purposely engineered to support a team.[19]

A business/system continuity plan (BCP) is a plan to help certify that business/system processes can persist during a period of emergency or crisis. Such a plan contains the following elements:

- Analysis of organizational/structural threats and hazardous risks
- A record of the key tasks needed to hold the company operations flowing
- Easily located management contact information
- Clarification of where personnel should go if there is a calamitous event
- Information on data backups and company site backup

Recovery time objective (RTO) is the maximum preferred time span tolerable between an unanticipated failure and the resumption of normal operations and service levels. The RTO determines the point in time after a failure at which the consequences of the intermission become unacceptable. Defining the RTO should be done as a component of the business impact analysis (BIA) accompanied by the BCP. In conjunction with the RTO, the acceptable amount of data loss evaluated in time, known as the recovery point objective (RPO), is taken into consideration. An inclusive BCP would then comprise the RTO, RPO, and BIA, in the company of different strategies and workaround procedures.[19]

3.2.2.2. Sustainable Manufacturing and Decision Support System

Sustainable manufacturing's research area reaches over numerous disciplines, and the movement toward sustainability can occur only through

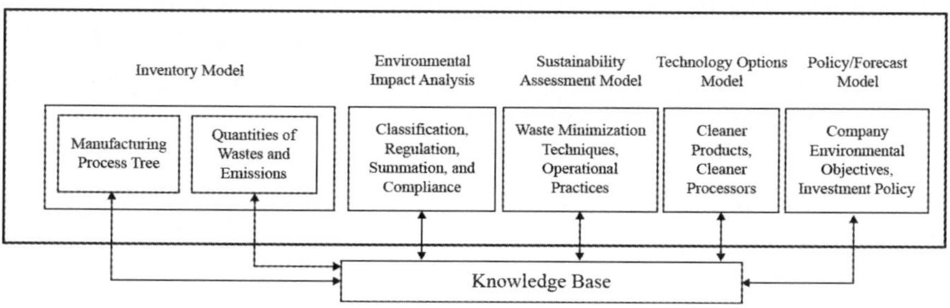

Figure 3.4. Sustainability Knowledge-Based Model Structure

transmissions, from behavioral to technological, and through holistic perspective along with local solutions. In addition, the demand for resource productivity in manufacturing is propelled by cost, protocols, and shareholder pressures. This matter is amplified by the complexity of activities and broader diversity among facilities. To cope with this issue, powerful DSS tools have been advanced to make possible the analysis of ever more interrelated and convoluted systems. In sustainable manufacturing, we attempt to use various assessment system tools in order to produce safer, cleaner, and environment-friendly products and services in such a way that minimizes natural resources at an affordable cost. To assess the sustainability level of a manufacturing organization, various computer-based DSSs such as fuzzy neural network, fuzzy group decision making, and multi-criteria decision making are developed and applied. Intelligent and knowledge-based DSS tools, which are designed for sustainable manufacturing, should support energy and material monitoring and recyclability and reuse analysis, along with other approaches to quantify the impact of products and processes on the environment. The structure of the sustainability knowledge-based model is presented through Figure 3.4.

Also, a sample of a decision tree approach with the aim of ranking and selecting the most suitable manufacturing technology is shown in Figure 3.5. This framework is developed for an experience-based manufacturing technology selection. It focuses on the relationship of historical decision cases for categorizing new manufacturing problems. The purpose of categorization is to order the most appropriate technology based on implementing cases in the past time horizon. With the help of categorization, successful technologies that are applied previously can be selected as potential options to be used in similar situations. Going through potential technologies and selecting alternatives with the highest possibility of success can result in distinguishing the most suitable solution or optimal solution for the problem. Hidden information patterns and knowledge, which are obtained by the information database analysis, are applied to support the categorization of the decision problem.

This methodology applies structured experience to select the most suitable technology on the basis of specified requirements for the defined decision variables. Experience is defined based on the data and information (a single expert or a group of experts would recall when assessing a new decision problem). This contains the performance of the selected technology, the project requirements that were formerly defined, and the success of the project. To represent a new manufacturing decision problem, the information is formed as a new decision case. The result of historical technology investment projects is obtained as a technology implementation case. Moreover, the feedback on the implementation testing of a technology is summarized to certify that lessons are learned of historical cases. The performance of previously implemented technologies provides essential feedback on the likelihood of the success of unknown decision problems.

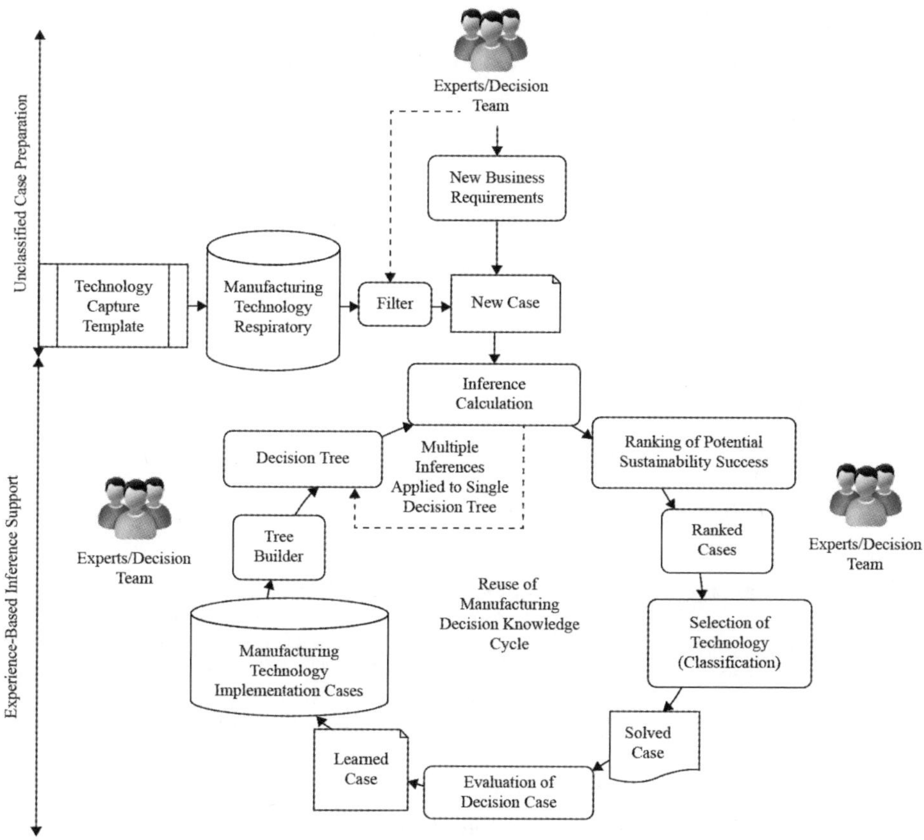

Figure 3.5. A Decision Tree Approach with the Aim of Ranking and Selecting the Most Suitable Manufacturing Technology Based on Previously Implemented Cases

Source: Evans, L., Lohse, N., and Summers, M. (2013), "A fuzzy-decision-tree approach for manufacturing technology selection exploiting experience-based information."

The success of a technology for a set of core problem requirements is conducted at the end of a project.[20]

Based on the EPA's definition of environmental tools and technologies for sustainable manufacturing, these kinds of tools and technologies are related to environmental protection, environmental assessment, compliance with environmental regulations, pollution control and prevention, waste management, remediation of contaminated property, design and operation of environmental infrastructure, and the provision and distribution of environmental resources. Many environmental tools and technologies can be used to green the manufacturing process and are therefore important to sustainable manufacturing. Examples of environmental tools and technologies include technologies for wastewater treatment, recycling, solid waste management, solar panels, wind turbines, and various sustainable electronic tools, databases, and toolkits introduced by the EPA, such as following sources:[21, 22]

- Electronics product environmental assessment tool (EPEAT)
- Energy star energy tracking tool (ETT)
- EPA chemical screening tool for exposures and environmental releases (ChemSTEER)
- EPA lean manufacturing and environment toolkits
- Sustainable facilities tool
- Industrial efficiency technology database by the Institute of Industrial Productivity
- Sustainable manufacturing toolkit by organisation for economic co-operation and development (OECD)
- Toxicity estimation software tool (TEST)
- Program for assisting the replacement of industrial solvents (PARIS)
- Waste reduction algorithm (WRA)
- Pollution prevention (P2) tool for metal finishing facilities by using computer-aided process engineering
- Tool for the reduction and assessment of chemical and other environmental impacts (TRACI)
- Database of sustainability indicators and indices (DOSII)

3.2.3. Scenario Development

One way to gain insights into an uncertain future of a system is to construct scenarios. This technique has been used to bring issues of environment and development to the attention of both scientists and policy makers. A scenario is described as a hypothetical sequence of events (based on a set of "if-then" propositions) built for the purpose of concentrating on casual processes and decision points (Fig. 3.6, Fig. 3.7).[23]

Scenario building goes through several steps. These begin with characterizing the current situation to define an issue or problem. Next, key dimensions, driving forces, strategic invariant elements, and crucial uncertainties are defined. These steps allow for assessing the logic of the system and, finally, developing images of the future regarding the studied system.

It is also important to distinguish the concepts and rationale behind the current system, and the recommended modifications originate from the scenario definition process. If a model-based methodology is applied for construction of a scenario, as is usually the case for environmental assessment, a conceptual model needs to be structured to recognize the key assumptions and decision factors, to bridge precise connections between the scenario definitions and the models to be used.[24]

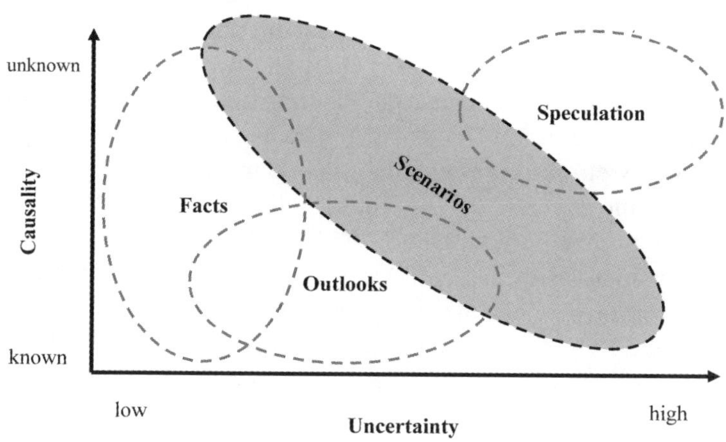

Figure 3.6. Developing a Scenario to Gain Insights into an Uncertain Future of a System

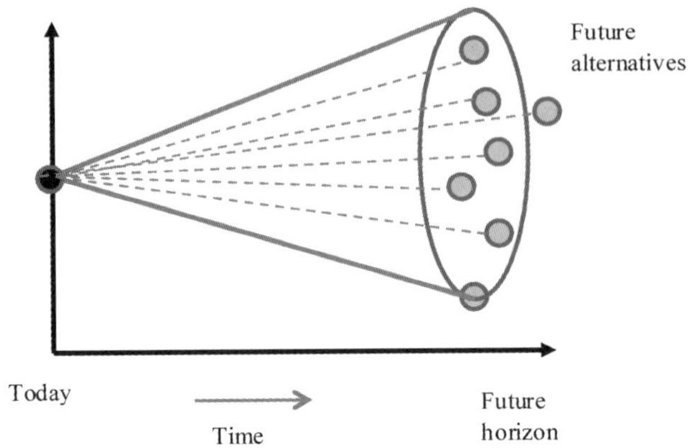

Figure 3.7. Conceptual Diagram of a Scenario Funnel

Scenario planning, also known as scenario thinking, is a strategic planning method applied to build flexible long-term plans. Scenario planning can benefit from computer simulations (specifically, discrete event simulation). It enables manufacturers to run their sustainability production models based on different conditions and assumptions that result in different hypothetical scenarios, which can be analyzed later to provide clues for making the best decisions.[25] The DES is a robust computing technique for exploring the behavior of systems. In plant and facility settings, DES software can be a fundamental tool in evaluating the efficiency. With the help of what-if scenarios, conclusions can be drawn on how to optimize a system's performance before its construction or exploitation. Even though DES is broadly employed as an analysis and problem-solving tool in factory designs, only recently has environmental sustainability been taken into account in combination with DES.

Life cycle assessment (LCA) is an analysis of a product's environmental impact throughout its life cycle. Life cycle inventory (LCI) as an element of the LCA comprises the data collection, modeling, description, and verification of data. The LCI database in general includes all pertinent data that assist to measure the material and energy flows into and out of the environment for related unit processes within study boundaries. Once the LCA is employed in the company of the DES, environmentally friendly

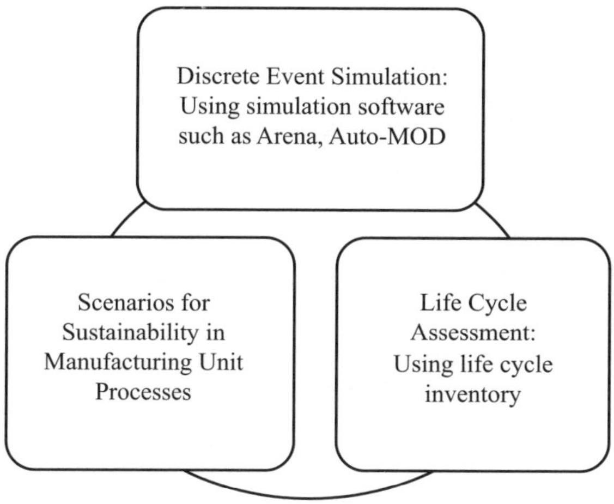

Figure 3.8. Integration of Discrete Event Simulation, Scenarios for Sustainability in Manufacturing Unit Processes, and Life Cycle Assessment

decisions can be made that lessen a manufacturing system's environmental impact on society. Combination of the DES, scenarios for sustainability in manufacturing processes, and the LCA can result in an efficient approach for improving sustainability in manufacturing system, as shown in Figure 3.8. Such an approach can assist manufacturing systems to analyze how shifting various parameters influence the waste materials and energy applied by the system along with the throughput.[26]

Generally, the scenario method for taking action regarding sustainability in manufacturing systems is composed of three phases:[27]

Phase one (Preparation)

- Developing an understanding of the system.
- Covering analysis of the interrelationships between the sustainability, production, costs, and benefits.
- Analyzing sustainability risks and mitigation measures.

Phase two (Scenario development)

- Developing a sustainability vision that mitigates or adapts to the risks identified in Phase one.

- Developing maps and forward (explorative) and backward (back casting) scenarios.
- The forward scenarios start from the current situation and recognize the successive technological and manufacturing system's changes that are likely in the near future based on causal relationships. The backward scenarios start from the vision and identify the preceding technological and manufacturing system changes necessary to reach the present state. The aligning paths are identified as the alternative innovation paths that the manufacturing system can follow toward system innovation.
- Incorporating current and future shareholders into the scenario map where they can be of high influence in achieving related goals.
- Generating technology, system, product, and service ideas that are located on the scenario map where they can be introduced if that specific state is reached in the future.

Phase three (Completion)

- Preparing an action plan or strategy, identifying the steps that need to be taken, and selecting proper assessment tools that are needed such as simulation software.

Moreover, it is critical to consider all elements of sustainable manufacturing (remanufacturing, redesigning, recovering, recycling, reusing, and reducing) in developed scenarios (Fig. 3.9).[28] Progressive phases of a scenario development are shown in Figure 3.10, and key questions to address throughout the scenario development for sustainability in manufacturing systems study may comprise:

- Important time horizon and intervals.
- Regional extent and subdivisions that should be considered.
- Manufacturing system components that should be considered in the scenarios.
- Including recycling, energy resources regulations, water resources regulations and policies, and so forth in the scenarios.
- Key assumptions about how different parts of the system work.

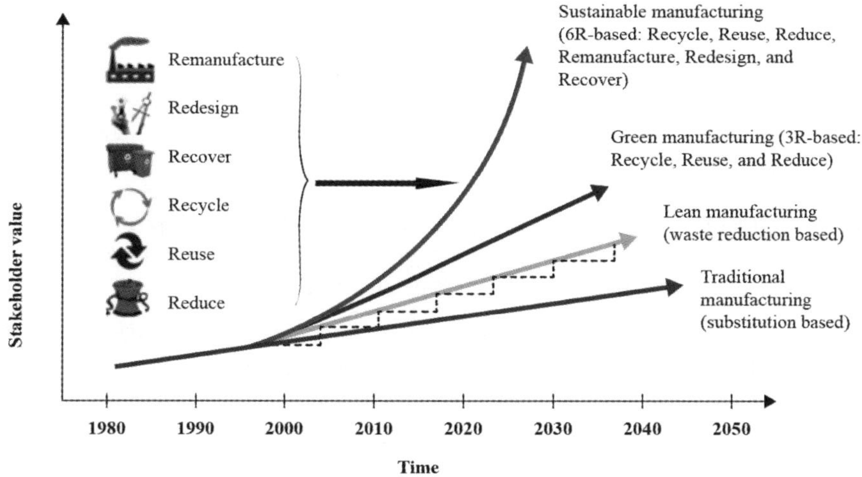

Figure 3.9. Evolution of Manufacturing Strategies

Source: Jawahir, I.S., Sustainable Manufacturing: The Driving Force for Innovative Products, Processes and Systems for Next Generation Manufacturing.[9]

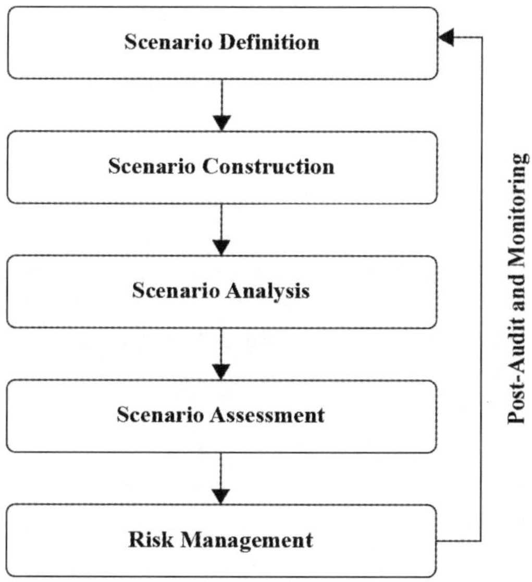

Figure 3.10. The Five Progressive Phases of Scenario Development

- Variables and situations that are important and the way they should be modeled.
- The spatial and temporal time-scales that are necessary for decision purposes.
- Sources of uncertainty needing to be considered.
- The magnitudes of the uncertainties and the way they propagate from one phase of a scenario development process into another. It refers to estimating uncertainty, which needs to be analyzed in scenario construction and scenario analysis.
- The way that uncertainties can be communicated to stakeholders and decision makers in the phases of scenario assessment and risk management.

Through evaluating and comparing yields from numerous scenarios, manufacturing systems can assess the sustainability impacts of their processes and products. They can shed light on some ways to improve their systems in terms of making them more sustainable without extra investment. On the basis of the suggestions, suitable measures can be done to get higher eco-efficiency by lessening idle time, optimizing the resource utilization, minimizing setup time, and leveling the line. A decision assistance management system can be associated with the simulation tools and defined scenarios; optimal solutions can be achieved in accordance with the decision makers' priorities and interests. Some sample outputs achieved from a general manufacturing simulation contain throughput, lead time, buffer levels, machine utilization, bottlenecks of the production line, and costs; while in simulation of sustainable manufacturing additional output measures such as energy consumption, carbon emissions, toxic material used, air pollutant emissions, liquid waste, solid waste, and other environmental impact are taken into consideration.

3.2.4. Material Flow Analysis and Energy Analysis

Material flow analysis (MFA) is an assessment method that estimates the efficiency of exploit of materials applying information from material flow accounting. It is a prevalent and standardized methodology for computing the input and output material flows of a system and supports to recognize waste of resources and materials (if not, they would be overlooked in conventional economic monitoring systems).[29] Related to this concept, economy-wide MFA and material flow accounts, developed by Eurostat, are the most standardized tools, which are chiefly applied at the national level with the likelihood of being used at other spatial levels.

3.2.4.1. The Material Balance Principle and Classifications of Materials in a System

The first rule of thermodynamics on the conservation of substance states that substance, that is, mass or energy, is neither created nor destroyed by any physical transformation (production or consumption) process. This material balance principle offers a logical base for the consistent and comprehensive recording of inputs, outputs, and material accumulation. The material balance theory can be used from either a systems viewpoint or a flow outlook.

For an assumed system, such as production or consumption processes, manufacturing systems, and companies, the material balance principle leads to the following identity:

Total inputs = Total outputs + Net accumulation, meaning that what substance enters the system is either accumulated in the system or leaving the system again as an output.

For an assumed physical flow, the material balance identity can be expressed as: Origin ≡ destination (other terms used are supply ≡ demand, or resources ≡ uses), denoting that all flows have a source and a destination, and a breakdown of the source must be far-reaching in the sense that the sum of the masses of the source must be equivalent to the sum of the masses by the destination.

Substance transforms during production and consumption processes. When this identity is applied to build economy-wide balances for particular material groups, for example, fossil fuels or biomass, the raw materials should be associated with, for example, the emissions or wastes that are the final destinations of these materials.

Classification of Material Inputs: The classifications of material inputs in a manufacturing system should allow distinguishing between renewable materials and nonrenewable materials. In such classifications, it is helpful to consider the following clustering:[29, 30, 31]

- Direct (or used) material inputs (solid, liquid, and gaseous materials) applied in production or consumption processes.

- Fossil fuels, minerals (metal ores, other industrial minerals, construction minerals), and biomass.
- Raw materials, semi-manufactured products, and finished products.
- Air (as in oxygen) and water demand estimation for some processes such as combustion.

Classification of Material Outputs: Material outputs of a manufacturing system can be classified into outputs to the environment and products. Outputs to the environment can be described as material flows flowing into the environment, either during or after production or consumption processes. Outputs to the environment can be categorized further into processed and unprocessed outputs. Unprocessed outputs stand for the disposal of unused extraction. Unused extraction refers to the materials that are moved or extracted from the environment throughout production activities but which were not applied in production or consumption processes (e.g., mining overburden, soil and rock excavated during construction and not used elsewhere, and harvest residues). Processed outputs are the result of production or consumption processes. They are categorized into various groups, such as emissions (emissions to air CO_2, SO_2, CO, dust, and NH_3 and emissions to water), waste flows (waste landfilled), dissipative use of products, and dissipative losses of materials. Dissipative uses of products and dissipative losses are described as the quantity (weight) of materials that are dispersed into the environment as a deliberate or unavoidable (with current technology) result of product use. Some examples of dissipative uses are fertilizers and manure spread on fields or salt and other thawing materials spread on roads. Some examples of dissipative losses are rubber worn away from car tires, particles worn from friction products such as brakes, abrasion from roads, losses from evaporation of, for example, water or other solvents carrying paints or other coatings. Generally, products can be considered as the following clusters:[29, 30, 31]

- Raw materials
- Semi-manufactured products
- Finished products
- Packaging material exported/shipped with products
- Waste exported for final treatment and disposal

Classification of Material Stocks: Stocks in the context of MFA are classified in various categories such as infrastructures, buildings, machinery, vehicles, durable goods, and so forth. Additions to the stock of infrastructures and buildings can be the construction materials for new constructions or renewal; removals can be construction and demolition wastes (and dissipative losses by corrosion and abrasion of infrastructures and buildings).

Related to machinery, vehicles, and associated durable inventories and products, additions can be new machines or new parts; removals can be the wastes from removing these durable goods. Demolition waste recycled and applied to new constructions must be incorporated under both gross additions and removals, leaving the net additions unaffected.[29, 30, 31]

3.2.4.2. Energy Analysis

Energy analysis can also be done by applying various kinds of energy measures, such as "EXERGY" and "EMERGY." Both of these assessments are more developed as they take both the quality and the quantity of energy into consideration. The "EXERGY" of a system is the highest amount of mechanical work that can be extracted, and "EXERGY" analysis provides an overview of the efficiency of resource exploitation and identifies where losses happen and where technological enhancements can be done to improve energy efficiency. In "EMERGY" analysis, all resources and products are expressed in common units (solar emjoules), and the needed solar energy for making them is computed. The presented aggregated framework in Figure 3.11 identifies energy consumption as an example metric by dividing the selected metrics to different levels throughout a manufacturing system.[32]

A hierarchal structure of a process metrics for energy consumption is demonstrated by Figure 3.12. The identification of sustainability pathways in a manufacturing system with consideration of both recycling materials and recovering energy is also shown in Figure 3.13.

3.2.4.3. Introducing an Integrated Software for Material Flow Analysis and Energy Analysis

The software of substance flow analysis (STAN) is a user-friendly freeware that helps performing the material flow analysis (MFA) based on the Austrian standard ÖNORM S 2096 (MFA—Application in waste management) under consideration of data uncertainties.

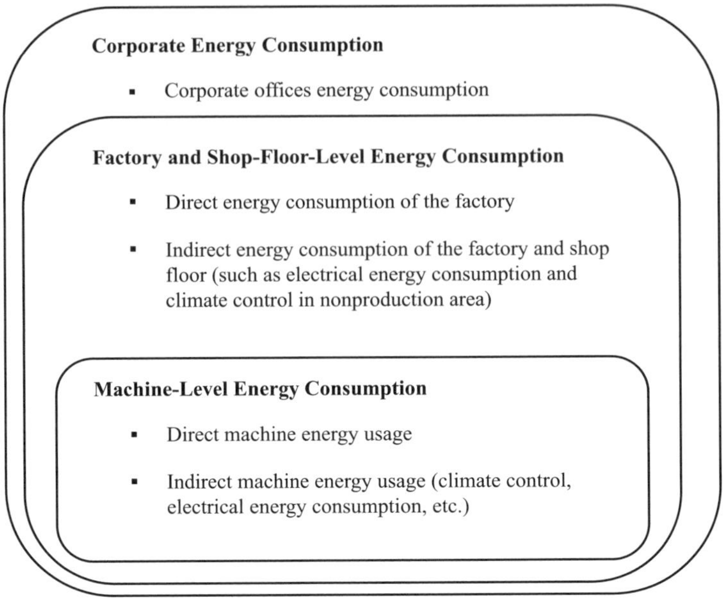

Figure 3.11. An Aggregated Framework for Identifying Energy Consumption Throughout a Manufacturing System

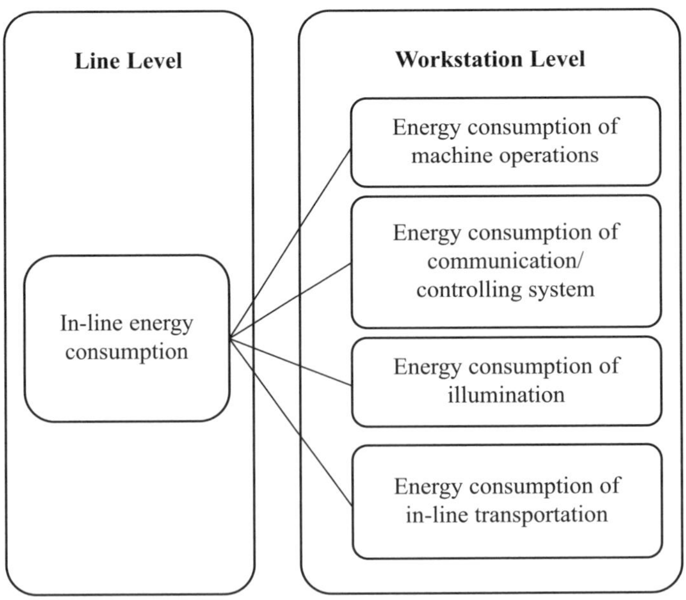

Figure 3.12. A Hierarchal Structure of a Process Metrics for Energy Consumption

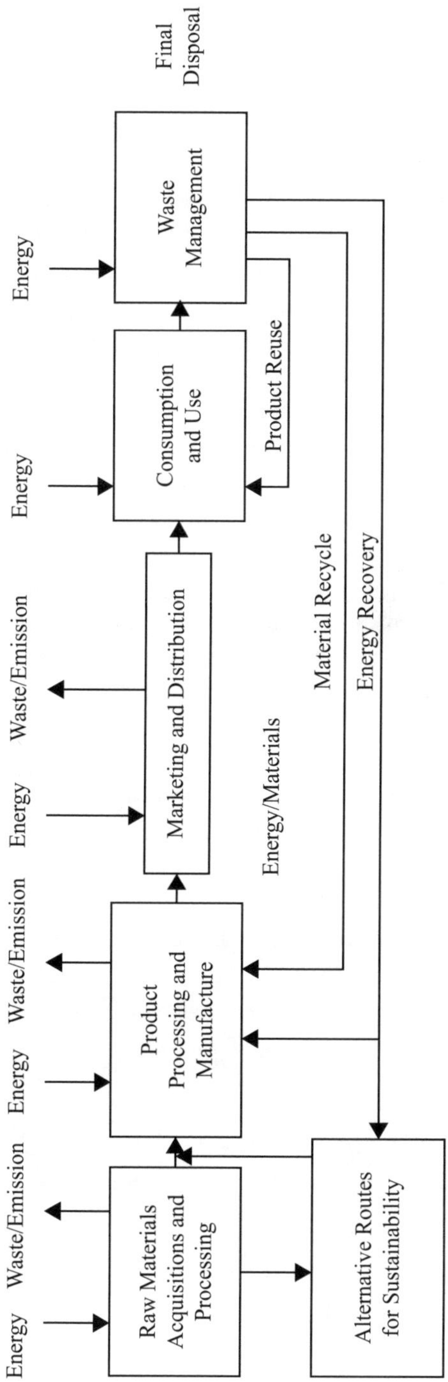

Figure 3.13. Examples of Process Loops Identifying Sustainability Pathways

Once a graphical model with predefined components (flows, processes, subsystems, system boundaries, and text fields) is constructed, known data (mass flows, stocks, concentrations, and transfer coefficients) with corresponding physical units can be either entered or imported for various hierarchical layers (product, substance, energy) and time periods. A graph of the model with flows demonstrated as specific arrows (the width of a flow is proportional to its value) can be printed or exported into different graphical formats. If adequate data are given, computation algorithms can be applied to reconcile uncertain data and to estimate unknown variables, including their uncertainties (error propagation). Statistical tests are also applied to reveal gross errors in a given data set. Since October 2006, STAN with different versions is available online for free download[33, 34] The language of the user interface containing the context sensitive help file can be changed between English and German. Simple and complex problems can be assessed by using this software. Here a simple system that can be analyzed through this software is presented. A layout for different elements in this given problem is also shown in Figure 3.14.

Assumption: Within a region products are created from imported raw materials and recycling materials. The products are sold and used within a year. A part of the products is reused in the production, and the rest is exported from the region as wastes.

Before starting a calculation to compute unknown quantities, the graphical model created with STAN has to be automatically translated into a mathematical model using four basic types of equations: balance equation (Eq. 23), transfer coefficient equation (Eq. 24), stock equation (Eq. 25), and concentration equation (Eq. 26). These equations could contain unknown and/or measured and exactly known (constant) variables.

$$\sum inputs = \sum outputs + change\, in\, stock \tag{23}$$

$$output_x = transfer\, coefficient_{to\ output\ x} \times \sum inputs \tag{24}$$

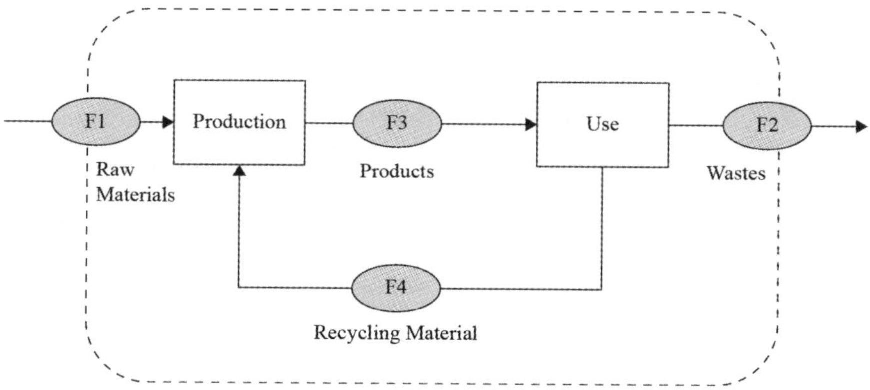

Figure 3.14. Default System (a Layout for Different Elements in the Given Problem)

$$stock_{period\ i+1} = stock_{period\ i} + change\ in\ stock_{period\ i} \quad (25)$$

$$mass_{substance} = mass_{good} \times concentration_{substance} \quad (26)$$

Calculate all flows for the following input data:

Question 1. Raw Materials = 20 (mass flow unit), Recycling Rate = 0.70

Solution: All unknown flows can be calculated. Because there is no stock within the system, the waste export must be the same size as the raw material import (Fig. 3.15).

Question 2. Raw Materials = 20 ± 2 (mass flow unit),Recycling Rate = 0.70

Solution: All unknown flows can be calculated. The uncertainty of the raw materials import influences all of the unknown flows (Fig. 3.16). All flows have the same relative uncertainty (10%).

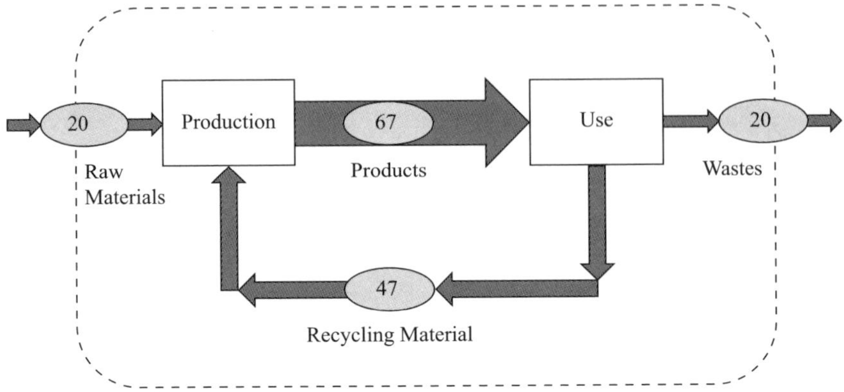

Figure 3.15. Solution for Q1

Question 3. Raw Materials = 20 ± 2 (mass flow unit), Recycling Rate = 0.70 ± 0.07

Solution: All unknown flows can be calculated as in Figure 3.17. Here the mathematical system is nonlinear and there are uncertain flows and uncertain transfer coefficients.

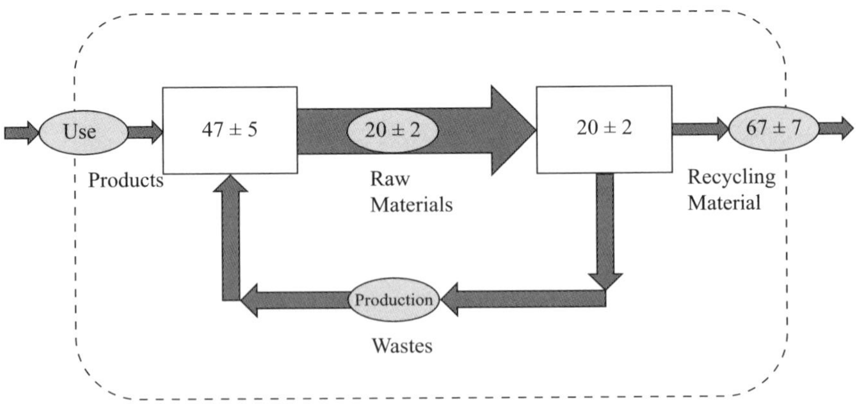

Figure 3.16. Solution for Q2

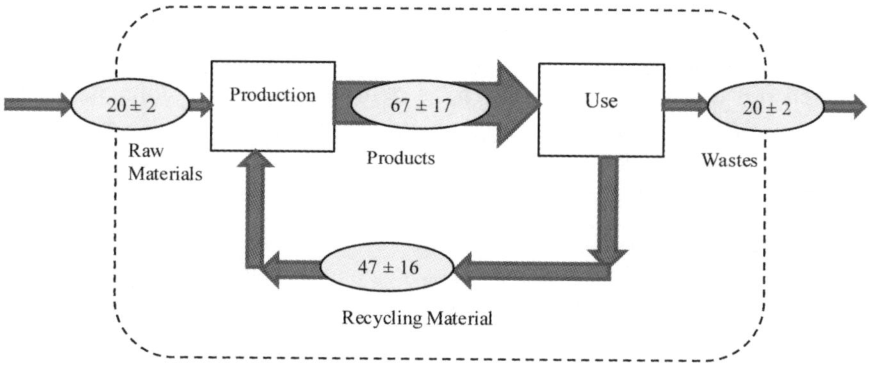

Figure 3.17. Solution for Q3

3.2.5. Life Cycle Assessment

Life cycle assessment (LCA) is a method to evaluate environmental impacts integrated with all the phases of a product's life (i.e., from raw material extraction throughout materials processing, manufacture, distribution, use, repair, maintenance, recycling, and disposal). An LCA can aid in avoiding a narrow viewpoint on environmental concerns through compiling an inventory of related energy and material inputs and environmental releases, analyzing the potential impacts related to the identified inputs and releases, and interpreting the outcomes and results to assist in making a more informed decision. ISO 14044, the standard that covers the LCA process, provides two major objectives, as follows:

1. First objective: analyzing the results, achieving conclusions, describing limitations, and providing recommendations in accordance with the findings of the preceding phases of the LCA, and reporting the results of the life cycle interpretation in a transparent manner.
2. Second objective: providing a speedily understandable, thorough, and consistent presentation of the outcomes of an LCA study, based on the goal and scope of the study.

3.2.5.1. Four Phases of the Life Cycle Assessment Process

The procedure of an LCA is well described by ISO 14040 and 14044 environmental management standards. Based on these standards, an LCA is carried out in four distinct stages, as demonstrated in Figure 3.18. The stages are often interrelated, as the results of one stage will advise how other stages are completed.[35, 36]

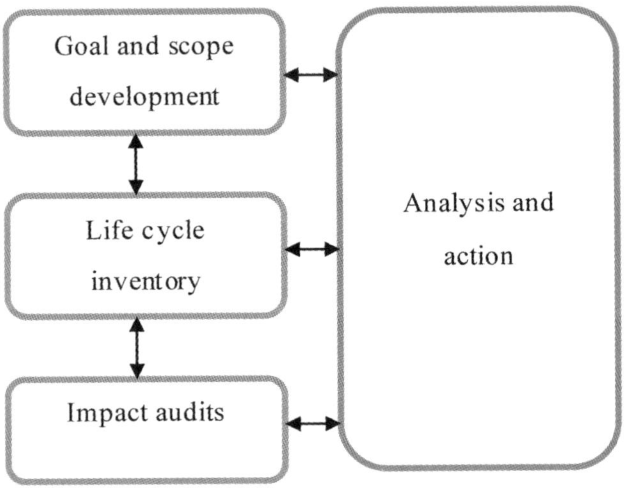

Figure 3.18. Four Main Phases of Life Cycle Assessment

Goal and Scope Development: This first phase includes clearing up the aim and extent of the LCA. It involves officially defining the functional unit, impacts of interest, and system boundary. This is a chief step, and the ISO standards necessitate that the objective and scope of an LCA be obviously defined and coherent with the intended application. Hence, the objective and scope documents entail technical features and details that guide succeeding work, such as:

- The functional unit that determines what exactly is being studied and measures the service delivered by the product system, giving a reference to which the inputs and outputs can be connected. The functional unit is an essential base that makes it possible to compare and analyze alternative products or services based on system boundaries, defined assumptions, and limitations.

- The allocation methods that are applied to divide the environmental load of a process when various products or functions share the same process.

Life Cycle Inventory: Life cycle inventory (LCI) assessment includes establishing an inventory of flows from and to the environment for a production system. Inventory flows contain inputs of water, energy, and raw materials and releases to air, land, and water. An LCI encompasses information on every environmental input and output related to a product or service (material and energy requirements, in addition to emissions and wastes). The inventory process looks simple enough in theory. In practice, nevertheless, it is dependent on several practical factors, such as geographical variations, data quality, and choice of technology. LCI has three chief phases:

1. Building a process flowchart that demonstrates raw materials, manufacturing processes, transports, uses, and waste management.
2. Collecting data for material inputs, products and by-products, solid waste, and air and water emissions.
3. Calculating the quantity of each kind of data corresponding to the functional unit. Fundamentally, this is the process flow diagram (with thorough information related to the mass and energy values).

The resulting LCI gives a breakdown of all of the energy and materials involved in a product's system at a level of detail that provides a foundation for assessment. In other words, to develop the inventory, a flow model of the technical system is built applying data on inputs and outputs. The flow model is normally exemplified with a flowchart that contains the tasks that are going to be analyzed in the pertinent supply chain and offers a pure image of the technical system boundaries. The input and output data necessitated for building the model are gathered for all tasks within the system boundary.

Moreover, the data should be associated with the functional unit determined by the objective and scope definition and can be displayed by tables with related interpretations.

Impact Audits: Impact audits are subsequent to inventory analysis. This LCA stage focuses on analyzing the importance of potential environmental impacts in accordance with the LCI flow outcomes. A classical life cycle impact audit contains:

- Choosing impact categories, category indicators, and characterization models.
- Classifying the inventory parameters and allocating them to specific impact categories.
- Measuring impact, where the categorized life cycle inventory flows are distinguished, using one of several possible impact audit methods, into common correspondence units that are then summed to present an overall impact category total.

In order to make it clearer, impact assessment consists of four specific steps:

- Classification: All kinds of materials are sorted into classes based on their influence on the environment.
- Characterization: All kinds of materials are multiplied by a factor that reveals their relative contribution to the environmental impact.
- Normalization: A comparison between the quantified impact and a certain reference value (for instance, the average environmental impact of a European citizen in one year) will be done.
- Weighting: Various value choices are given to impact categories to create a single score.

For each type of substance, a schematic cause-response route can be advanced that explains the environmental mechanism of the substance emitted. Besides this environmental mechanism, an impact category indicator outcome can be selected either at the midpoint or endpoint level.

- Midpoint impact category, also known as the problem-oriented methodology, translates impacts into environmental themes such as climate change and acidification.
- Endpoint impact category, also known as the damage-oriented methodology, translates environmental impacts into issues of concern, such as human health, natural environment, and natural resources. Endpoint out

comes, which have a higher level of uncertainty in comparison with midpoint outcomes, nonetheless are simpler to be comprehended by decision makers.

Analysis and Action: This stage is a methodical procedure to recognize, quantify, check, and analyze information from the outcomes of the life cycle inventory and/or the life cycle impact assessment. The outcomes from the inventory analysis and impact assessment are reviewed throughout the analysis and action stage. The result of the analysis and action stage is a set of conclusions and suggestions for the related study. Based on ISO 14040, the analysis and action should entail the following:

1. Identification of major issues in accordance with the outcomes of the LCI and impact audit stages of an LCA.
2. Assessment of the study (in view of completeness, sensitivity, and consistency checks).
3. Conclusions, restrictions and suggestions. (A great deal of an LCA's value depends on what way its outcomes are communicated to designers, engineers, managers, marketers, customers, people, or other parts of the supply chain involved in making related decisions.)

A key aim of performing life cycle analysis and action is to define the level of confidence in the ultimate outcomes and convey them in a fair, complete, and accurate way. Elucidating the outcomes of an LCA begins with understanding the precision of the outcomes and being assured that they satisfy the objectives of the related study. This is achieved by recognizing the data components that contribute notably to each impact category, assessing the sensitivity of these major data components, analyzing the consistency and completeness of the related study, and extracting conclusions and suggestions on the basis of a clear image of how the LCA was performed and the outcomes were developed. Based on ISO 14044, the related interpretations explain numerous checks, such as checking the level of uncertainty and conducting sensitivity analysis and contribution analysis, that we need to make to be sure the conclusions are sufficiently supported by the data and procedures applied in the related study.

3.2.5.2. Life Cycle Assessment and Sustainable Manufacturing

Sustainable manufacturing is a systematic approach to the formation and distribution of innovative products and services that optimizes resources' use (resources: materials, energy, water, and land), eliminates toxic materials, and aims ideally at achieving a zero-waste production system that reduces carbon footprints across the whole life cycle of products and services. Suitable information infrastructure and tools should be developed to be able to cover the total life cycle synthesis and sustainability into the early design stage of products. Current LCA tools are chiefly appropriate for post-design, production, and distribution, while in order to do an assessment and synthesis in the early design phase, it is vital to incorporate environmental aspects into product design and development, product life cycle management, and the total life cycle analysis. One of the important concepts in sustainable manufacturing is the application of the 6R methodology (recycle, reuse, reduce, remanufacture, redesign, and recover). This methodology promotes the total product life cycle approach to enable the ideal case of achieving closed-loop material flow through multiple product life cycles, as illustrated in Figure 3.19.[37] Moreover, in sustainable manufacturing systems modeling (typically static simulation modeling) life cycle inventory can be applied to calculate the quantities of various resources required and emissions and waste generated per functional unit.

A combination of value stream mapping (VSM), discrete event simulation (DES), and life cycle assessment (LCA) can provide manufacturers with a more powerful tool in the sustainability journey. It aids engineers in speeding up the design process and improving decision making.

VSM is considered a lean tool for enhancing efficiency by recognizing wasteful material, energy, and information flows. We will describe VSM as one of the lean manufacturing methodologies in section 3.4.3. ("Lean Manufacturing: Methodologies and Their Application in Process Optimization") of this chapter. VSM is an easy-to-use symbolic process modeling technique. It specifies the activities, cycle times, downtimes, and delays, and distinguishes bottlenecks and non-value-added tasks in the production or logistics systems. As mentioned earlier, DES is applied to model sequence of operations in the manufacturing systems, and it provides dynamic analysis. Through the integration of simulation with VSM, it is possible to add some related parameters to analyze bottlenecks, buffer levels, and changes in production volumes. Simulation helps in adding parameters, modeling scenarios, and uncertainties. Solding and Gullander

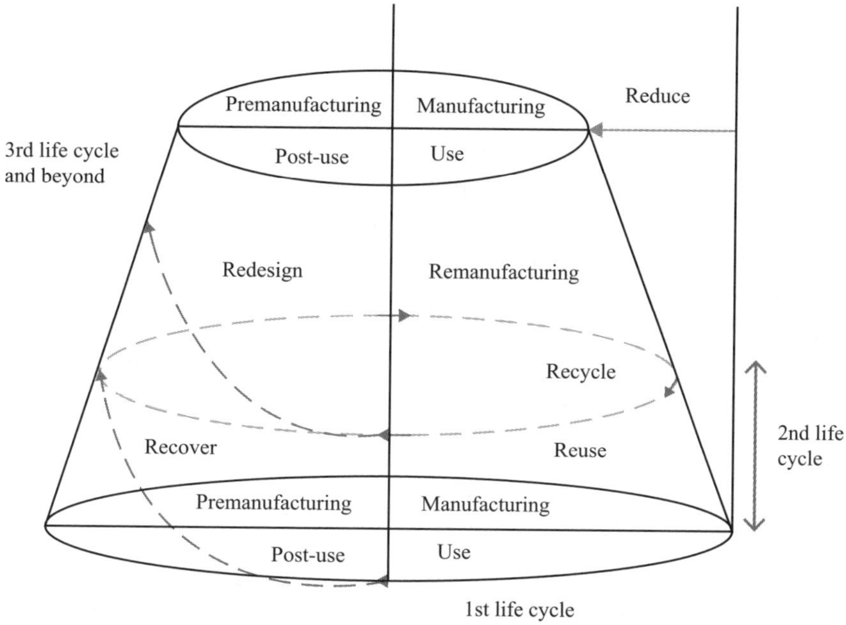

Figure 3.19. Applying 6R Methodology Throughout the Total Product Life Cycle

proposed a simulation-based VSM that made use of AutoMod and Excel. They showed that simulation can make the VSM more dynamic and efficient.[38] There are various ways to automate the creation of a simulation model from the digital VSM model. A particular software tool, for instance, Process Simulator, Simul8, and VisioSim support simulation of the process maps created with Microsoft Visio, whereas, others, such as Arena, SimCad, and Extend offer specific VSM templates. The Connecticut Centre for Advanced Technology, Inc. (CCAT), in 2009, verified the capacity of semi-automatically converting VSM models created with Visio to QUEST simulation models. The CCAT used the core manufacturing simulation data (CMSD), which is developed under the Simulation Interoperability Standards Organization (SISO).

LCA is also integrated with a simulation model in the simulation tool for joint assessment of ergonomics, level of automation, and environmental impacts (SIMTER). SIMTER is a holistic, integrated production system simulation and environmental analysis method. It plays the role of a cross-disciplinary decision-making technique for sustainable and concurrent analysis of the level of automation, human ergonomics, productivity, and environmental aspects in the manufacturing system. In SIMTER, the selected environmental information was taken from the European Commission's public life cycle inventory (LCI) database. The SIMTER development could be improved with other commercial simulation software. Another example is the newest version of Simul8 DES software, which provides carbon monitoring. VSM and LCA have also been combined together with several applications. The U.S. Environmental Protection Agency (EPA) has introduced the Lean and Environment Toolkit as well. This toolkit is prepared for the lean specialist to enhance his or her environmental performance in manufacturing and production systems. It concentrates on removing wasteful operations by classifying tasks in value-added and nonvalue-added activities and hence can be applied to smooth environmental improvements.

Typically, methods and tools for environmental analysis in the discrete-part manufacturing industry can be categorized into two classes: those that are applied chiefly for the assessment of environmental aspects, such as LCA, and those that are applied for environmental analysis as add-on tools. These tools need less effort to be modified with existing manufacturing modeling tools, although they compromise on comprehensiveness in the environmental assessment part. With the integration of LCA, VSM, and DES, DES plays the role of an add-on tool and VSM denotes a visualization technique applied to implement environmental indicators. These techniques have various functions, but they can complete with one another, as shown in Table 3.4.

Life cycle costing (LCC) also gives information on costs during the entire life cycle of a product or part. Conventional LCC is a technique of computing the total cost of a product or services generated during its entire life cycle, from its purchase to its disposal, consisting of design, installation, operation, maintenance, recycling, and disposal. It is better to implement a thorough cost analysis frequently in sustainable manufacturing for decision making, as sustainable manufacturing–oriented modifications either result in a cost saving or might need additional cost of operations. Decisions should be made in accordance with the cost and payback of the activities before implementing the sustainability programs. Current standards and guidelines describe LCC as the consideration of all costs that a

Table 3.4. Characteristics of VSM, LCA, and DES with Preferred Features Highlighted in Italics

Feature	**VSM**	**LCA**	**DES**
Dynamic assessment	Deterministic, standard, or average parameter	Deterministic, standard, or average parameter	*Dynamic event relationships; probabilistic parameters*
Publicly available data		*Public LCA data available*	
Visualization	2-D process map	Limited process view	*3D visualization and animation*
Simplified	*User-friendly tool*	Expert tool	Expert tool
Standardized	Industrial de facto standard for lean manufacturing	*Standardized ISO14040, ISO 14044*	Partially
Framework for environmental impact analysis	Methodology has been presented	*The main tool*	Mostly research initiatives; also commercial solution entering the market

system causes over its lifetime. The cost can be classified into the stages of "before use," "during use," and "after use."[39, 40] The major goal of using LCC is to optimize the cost of acquiring, owning, and operating physical assets over their useful lifetime by recognizing and measuring all the substantial costs integrated with that lifetime, applying the present-value technique.

Environmental LCC advances conventional LCC such that it evaluates all costs integrated with a product's life cycle that are covered by one or more of the actors in the product's life cycle. These actors are suppliers, manufacturers, customers, end-users or end-of-life actors.[40, 41] A variation of interest can happen in terms of the optimization of costs and revenues in the product life cycle. Activities deriving from this variation are the assessment of the LCC and profits based on the distribution of costs within the three stages of "manufacturing," "usage and service," and "recycling and reuse." The manufacturing stage contains costs for engineering, material, manufacturing, sales, and procurement in addition to a benefit defined by the purchase price, and it is clear that the value of products is increased throughout the manufacturing stage. A first calculation is conducted by fixing the price. After starting the usage phase, there is a loss in the value of products because of wear and attrition. In this stage, the total value can be summed up as demonstrated by Figure 3.20. Losses in wear and attrition are to reactivate as a result of upgrading and service.

$$value = \frac{desired\ performance\ or\ function\,(F)}{overall\ costs\,(C)} \tag{27}$$

Increasing value through $\frac{F\uparrow}{C};\ \frac{F\uparrow}{C\downarrow};\ \frac{F\uparrow\uparrow}{C\downarrow};$ or $\frac{F}{C\downarrow}$

The LCC process can be as easy as a table of predictable annual costs, or it can be a complex model that enables to create various scenarios on the basis of the assumptions about upcoming cost drivers. The range and complexity of the LCC analysis should normally reveal the complexity of the assets under investigation, the ability to forecast future costs, and the importance of the future costs to the decision being decided by the manufacturing system.

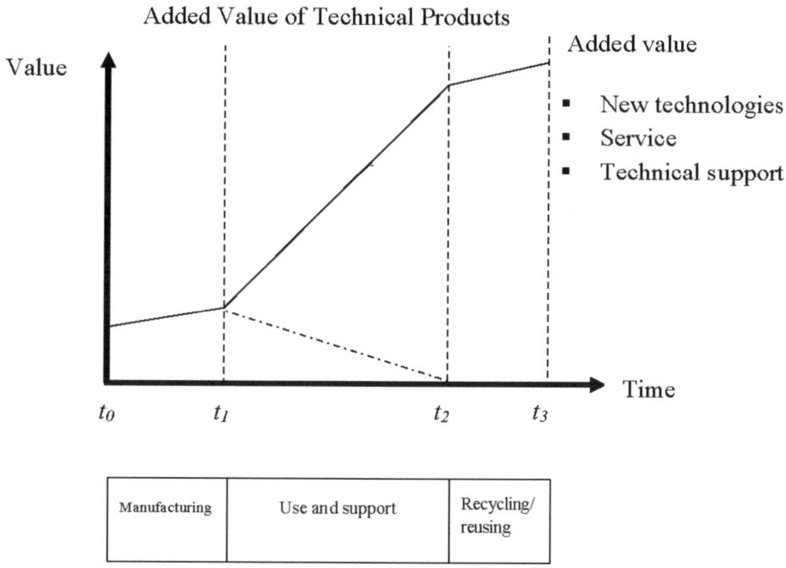

Figure 3.20. Stages of Value Creation in the Product Life Cycle

An LCC assessment contains the analysis of the costs of a system or a component over its entire life span. The typical costs of a system might entail:

a) Acquisition costs (or design and development costs)

b) Operating costs: cost of failures, cost of repairs, cost for spares, downtime costs, cost of loss of production

c) Maintenance costs: cost of corrective maintenance, preventive maintenance, predictive maintenance

d) Disposal costs

The working group of the European Society of Environmental Toxicology and Chemistry considers LCC not only as a cost calculation method but also as a controlling method for the estimation of all costs associated with an artifact. The group distinguishes between internal and external costs. Internal costs are all of the cash flows of the actors involved within the economic system. External costs are compensations for the product's impact on the social and ecological environment. They proposed to mainly work with internal or already internalized costs of taxes or subsidies.[42]

A thorough life cycle cost projection (LCCP) assessment might also entail other costs, as well as other accounting/financial components (such as interest rates, depreciation, and present value of money/discount rates). The life cycle cost of an asset in a system can be explained by this formula:

life cycle cost = initial (projected) capital costs + projected lifetime operating costs + projected lifetime maintenance costs + projected capital rehabilitation costs + projected disposal costs – projected residual value

The following equation is also the general formula for the LCC present-value model:

$$LCC = \sum_{t=0}^{N} \frac{C_t}{(1+d)^t} \tag{28}$$

Where

LCC: Total LCC in present-value dollars for a given alternative.

C_t: The sum of all relevant costs, including initial and future costs, less any positive cash flow, occurring in year t.

N: Number of years in the study period.

d: Discount rate applied to modify cash flows to present value.

Moreover, in order to evaluate the net saving (NS), a measure should be applied. Net saving measure is a change of the net benefits (NB) measure of economic performance of a project such as a defined sustainability project for a manufacturing system. The NB method quantifies the variance between present-value benefits and present-value costs of a specific investment over the designed study period. The NB measure is typically used when positive cash flows are planned to validate the investment in a project. The NS method is used when benefits happen chiefly in the form of future operational cost reductions (for instance, energy and water cost savings). The NS technique computes the net amount (in present-value dollars) that a project alternative is anticipated to save over and above the amount that would have been earned from investing the same funds at the minimum acceptable rate of return (i.e., the discount rate). The NS for a project alternative, relative to an assigned base case, can be computed simply by subtracting the LCC of the alternative from the LCC of the base case. That is:

$$NS = LCC_{base\ case} - LCC_{alternative} \tag{29}$$

Providing that the NS is greater than zero, the project is considered to be cost effective relative to the base case. This is similar to requiring that the LCC of a project alternative is lower than the LCC of its base case. When evaluating multiple, mutually exclusive project alternatives, the alternative with the greatest NS will be the similar alternative that has the lowest LCC. Hence, the LCC and NS techniques are completely consistent and can be applied interchangeably. The benefit of the LCC method relative to the NS method when assessing multiple alternatives is that the former does not need that the base case be definitely identified. NS can also be computed from individual cost differences between the base case and alternatives (e.g., differences between initial investment costs, between energy costs, and between operation management and research costs). Net savings can be computed applying individual cost changes by using the following typical formula:

$$NS_{A:BC} = \sum_{t=0}^{N} \frac{S_t}{(1+d)^t} - \sum_{t=0}^{N} \frac{\Delta I_t}{(1+d)^t} \tag{30}$$

Where:

$NS_{A:BC}$: NS, in present-value dollars, of the alternative (A), relative to the Base Case (BC)

S_t: Savings in year t in operational costs related to the alternative

ΔI_t: Additional investment-related costs in year t related to the alternative

t: Year of occurrence (where 0 is the base date)

d: discount rate

N: Number of years in study period

(Please consider that once the summation index (t = 0 to N) is shown for operational savings, these savings will normally not be incurred on the base date but only after the project is put into service.)

Like LCC, *cost of ownership* is a frequently applied term in manufacturing in general and has applications in sustainable manufacturing. A complete economic analysis metric is applied for cost assessment and comparisons among different alternative options in sustainable manufacturing.[43] The cost elements that are usually taken into consideration in this analysis entail the equipment costs, setup costs, and annual costs for operation of facilities. The equipment costs are the purchasing costs of the equipment and related to the components; setup costs entail those for installation, transportation, and engineer training and also the labor cost for installing the facility on the site; annual operational costs consist of those for electricity, space consumed on the factory floor, maintenance, and consumables. As the equipment costs and setup costs are one-time capital costs happening in the initiation phase of the project, the annual operational costs reach into the future depending on the planned lifetime of the equipment for operation. To compare the cost of alternatives on the similar benchmark, typically specialists have the annual operational cost projected to the present value, on the basis of the economic discounting method. A thorough model for the cost of ownership is shown as follows:

$$cost\ of\ ownership = \frac{\left(C_1 + C_2 + \frac{C_3}{(1+i)^t}\right) \times N}{B \times V_E} \tag{31}$$

Where C_1 is equipment cost ($ in first year);

$C_2 = I + T + \Sigma\ (P \text{ x} F)$ ($ in first year);

I: Installation fee including labor cost ($);

T: Transportation fee ($);

P: Number of people to be trained;

F: Training fee for each person ($);

C_3 is annual operation cost and is calculated as Equation 32:

$$C_3 = S \times R_s + U \times R_E + \Sigma O + H \times R_M \tag{32}$$

S: Footprint for the equipment ($feet^2$);
R_S: Footprint cost rate ($\$/feet^2/year$);
U: Electricity consumed (kilowatt hour/year);
R_E: Electricity rate ($/kilowatt hour);
O: Cost of various consumables ($);
H: Downtime (hours);
R_M: Maintenance cost ($/hour);

N is the number of pieces of equipment needed for the system;
i is discounting rate;
t is t years into the future;
B is total amount of output from the system;
V_E is the economic value of unit output.

Various kinds of scenarios can be taken into consideration when we focus on the integration of LCA with LCC. Various types of scenarios can be categorized as follows:

- Predictive scenarios that try to answer this question: What will happen?
- Explorative scenarios that attempt to answer this question: What can happen?
- Normative scenarios that aim at answering this question: How can specific objectives be achieved?

Predictive scenarios are efficient tools, as a sample, for electricity production and other background processes in LCAs. If the time period is longer and the uncertainty in the estimations increases, it may be helpful to contain several explorative scenarios, which together describe possible future developments of, for instance, the energy systems. This can be considered as a particular type of uncertainty analysis. Various studies on waste management systems have, for instance, incorporated explorative scenarios in comparing fuels with solid waste as a fuel, an aspect that can be pivotal for the outcomes of waste management LCAs and LCCs. Explorative scenarios have also been applied in, for instance, energy and transportation studies. A normative scenario can be applied in LCA as an explorative scenario explaining what can occur.[44–47]

3.2.6. Risk Assessment

Risk management and risk assessment have critical roles in helping organizations and manufacturing systems to identify, quantify, and find ways to mitigate any risks associated with new sustainable products and practices. Risk assessment is the process of summarizing a scientific examination of a specific risk (what are the probability and the severity of the adverse outcome?). Risk management is the process of recognizing and employing measures that can be used to decrease risk to a tolerable level. Integration of risk assessment and risk management in a sustainable manufacturing system helps manufacturers to identify and quantify the risk and to select necessary action regarding their system and environmental status. Risk assessment is the process by which hazards are recognized and the level of risk is defined through quantitative or qualitative analysis. A reliable risk assessment methodology typically contains these stages:

1. Recognizing hazards and risks related to the work system or process
2. Measuring and evaluating occurrence of exposure and severity of consequence that cause a loss to occur, in addition to probability of occurrence
3. Analyzing risks related to the work system and deciding proper ways to manage hazards and reduce risks to an acceptable level
4. Developing and implementing additional mitigating controls, if necessary, to reduce risks to a tolerable level
5. Monitoring the effectiveness of controls and periodically observing to recognize potential new risk exposures

The objective of the risk assessment process, and succeeding controls, is to attain a tolerable level of risk. The risk assessment and mitigation processes are not complete until acceptable risk levels are gained.[49] Acceptability is a moving target. As knowledge about a specific risk increases, so does the possibility to make a more informed judgment of its acceptability. The risk that is determined to be an acceptable risk today may become an unacceptable risk in the near future. When the risk assessment process is applied as a continuous improvement process, residual risk can be diminished over time and deliver improvements in safety performance that may be sustained over time. Part of risk assessment can also entail financing the risks in addition to controlling the frequency and severity of risk. Moreover, it is worthwhile to mention that in developing improvement action plans for making processes more sustainable, some risks will be reduced but sometimes some other new risks will appear, and this is where insurance also comes into play.[50]

The World Business Council for Sustainable Development counts sustainability risks among the new megarisks of the century and notes that these risks have weighty meaning for enterprises, organizations, and manufacturing systems. It recommends applying a risk approach when organizing for sustainability in a manufacturing system.[51]

The EPA defines the risk as the chance of harmful effects to human health or to ecological systems stemming from exposure to an environmental stressor. In the EPA's definition, a stressor is any physical, chemical, or biological entity that can entail an adverse response. Stressors might adversely have effects on specific natural resources or on whole ecosystems.[52] A sample of risk categories in a sustainable manufacturing system is demonstrated in Table 3.5.

Risk is also defined as the probability of occurrence of an adverse result and the severity of the consequences if the outcome does happen. A straightforward approach for viewing risks in a sustainable manufacturing system concentrates on two vital aspects of a potentially disruptive event:

1. Probability (likelihood) of the event actually occurring
2. Impact (consequence) of the event on the sustainable manufacturing system, products, services, and environment

In general, risk analysis begins with risk identification. When all of the events that could potentially disrupt the sustainable manufacturing system are recognized, different techniques can be applied to measure the probability of risk occurrence and the related potential risk impact. Then efficient techniques for controlling the risks should be advanced. Efficient risk management needs measuring risks to position them in their appropriate context and to weigh the risk costs and profits of making specific decisions.

Stochastic formulation and simulation modeling (such as Monte Carlo methods, which are a wide category of computational algorithms that count on frequent random sampling to obtain numerical outcome and distribution

Table 3.5. A Sample of Risk Categories in a Sustainable Manufacturing System

Category	Examples
Operational/Technological	• Forecast errors • Life cycle analysis errors • Component/material storage • Capacity constraints in recycling materials • Quality problems in recycling and reusing materials • Machine failure/downtime • Software failure • Imperfect yields • Imperfect efficiency in recycling and recovering processes • Process/product changes • Property losses (owing to accidents, etc.) • Transportation risks (delays, damage from handling/ transportation, re-routing, etc.) • Increased green gas emissions of the related plants • Elevated energy consumption, storage risks (incomplete customer order, insufficient holding space, etc.) • Budget overrun (emergence of a disruptive technology) • Communication or IT disruptions
Social	• Labor shortages • Lack of experts and skillful staffs • Human errors, • Negative media coverage (reputation risk) • Decreased labor productivity
Natural/Hazardous	• Fire • Severe thunderstorm • Heat wave • Flood • Tornado • Hurricane
Economy/Competition	• Interest rate fluctuation • Exchange rate fluctuation • Commodity price fluctuation • Stock market collapse • Global economic recession • Bankruptcy of partners
Legal	• Liabilities • Law suits • Governmental incentives/restrictions • New regulations • Change in tax structures • Customs risks (inspection delay, missing data on documentation)

of an unknown probabilistic entity) provide general means to formally signify the uncertainties associated with risk events. Probabilistic modeling can also be utilized to assess the impact of variability and uncertainty in the different input parameters, for example, environmental exposure levels and transport processes. This technique applies the complete range of input data to advance a probability distribution of exposure or risk instead of a single point value. Manufacturing systems can also apply a risk root cause and sensitivity analysis to comprehend the drivers and pathways of system risks integrated with their sustainability program. Another effective tool that could be used successfully in senior sustainability management during the sustainability risk assessment process to raise awareness and increase visibility of risks is risk matrix, also known as probability and impact matrix (Table 3.6 and Table 3.7). A risk matrix is a tool to identify and rank process risks that are usually characterized through one or more multifunctional inspections (e.g., process hazard analysis, audit, or incident investigation). In some cases with the purpose of having more accurate analysis, it is recommended to develop a fuzzy risk matrix using fuzzy theory. In this method, the frequency and severity of the consequences of the incident scenario will define as fuzzy numbers. A fuzzy number is a generalization of a regular, real number in the sense that it does not refer to one single value, but rather to a linked set of possible values where each possible value has its own weight between 0 and 1. This weight is called the membership function. Fuzzy numbers are an expansion of real numbers. Computations with fuzzy numbers provide the integration of uncertainty with parameters, properties, geometry, initial conditions, and so forth.[53] Moreover, as shown in Figure 3.21, the Define, Measure, Analyze, Improve, and Control (DMAIC) approach can also be utilized in sustainability risk assessment.

Table 3.6. Risk Rating Matrix That Can Be Applied for Sustainability Risk Estimations (Risk Rating = Likelihood × Severity)

Severity (Impact)	Catastrophic	5	5	10	15	20	25
	Significant	4	4	8	12	16	20
	Moderate	3	3	6	9	12	15
	Low	2	2	4	6	8	10
	Negligible	1	1	2	3	4	5
			1	2	3	4	5
			Improbable	Remote	Occasional	Probable	Frequent
			Likelihood				

Table 3.7. Risk Matrix That Can Be Applied for Sustainability Risk Assessment

Risk Probability (Likelihood)							
	Frequent	5	5A	5B	5C	5D	5E
	Likely	4	4A	4B	4C	4D	4E
	Occasional	3	3A	3B	3C	3D	3E
	Seldom	2	2A	2B	2C	2D	2E
	Impossable	1	1A	1B	1C	1D	1E
			A Catastrophic	B Critical	C Moderate	D Minor	E Negligible
			Risk severity (Impact)				

Assessment Risk Index	Criteria
5A, 5B, 5C, 4A, 4B, 3A	Unacceptable under exsiting Circumstances; It requires immediate action by risk review board, quality and safety department, or sustainability department.
5D, 5E, 4C, 3B, 3C, 2A, 2B	Manageable under risk control and mitigation; Risk review board will take proper action and make essential decisions related to that.
4D, 4E, 3D, 2C, 1A, 1B	Acceptable after review of the operation; It requires continued tracking and recorded action plans.
3E, 2D, 2E,1C, 1D, 1E	Acceptable with continued data collection and trending for continuous improvement.

3.2.7. Environmental Impact Assessment

Environmental impact assessment (EIA) is a process of assessing the probable environmental impacts of a recommended project or development, considering interconnected socioeconomic, cultural, and human health impacts, both valuable and adverse. The United Nations Environment Programme (UNEP) describes the EIA as a tool applied to recognize the environmental, social, and economic impacts of a project before decision making. It attempts to forecast environmental impacts at initial stages in project planning and design, discover ways and means to reduce adverse

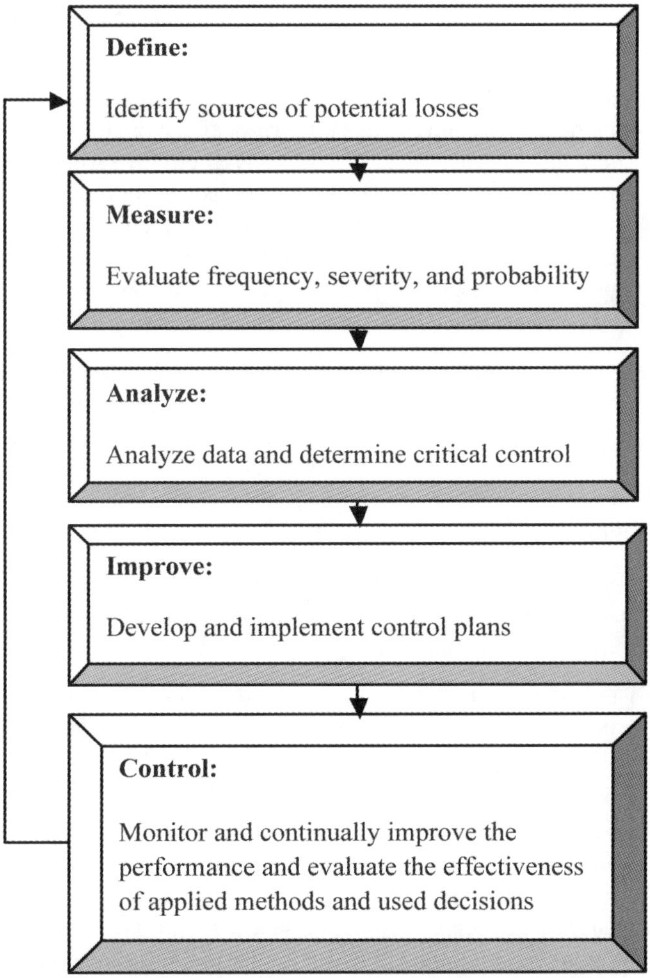

Figure 3.21. DMAIC Approach in Sustainability Risk Assessment

impacts, shape projects to suit the local environment, and present the forecasts and options to decision makers. With the application of an EIA, both environmental and economic benefits can be gained, such as reduced cost/time of project design and implementation and moderated treatment/cleanup costs. The key components of an EIA would essentially include the following steps:[54, 55, 56]

1. Screening to decide which projects or developments need a full or partial impact assessment study.
2. Scoping to recognize which potential impacts are related to assess (based on legislative requirements, international conventions, expert knowledge and public involvement), to identify alternative solutions that prevent, mitigate, or compensate adverse impacts on biodiversity (containing the option of not proceeding with the development, finding alternative designs or sites that prevent the impacts—including safeguards in the design of the project—or providing compensation for adverse impacts), and eventually to derive terms of reference for the impact analysis.
3. Assessment and evaluation of impacts and development of alternatives to forecast and recognize the likely environmental impacts of a proposed project or development, containing the detailed elaboration of alternatives; some techniques can be applied in forecasting the impacts, such as simulation, mathematical modeling, risk assessment, economic evaluations.
4. The Environmental Impact Statement (EIS) or EIA report: incorporating an environmental management plan (EMP) and a nontechnical summary for the general audience.
5. Review of the EIS according to the terms of reference (scoping) and public (including authority) participation.
6. Decision making on whether to approve the project or not, and under what conditions.
7. Monitoring, compliance, enforcement, and environmental auditing. Monitor whether the forecasted impacts and proposed mitigation measures happen as defined in the EMP; validate the compliance of the proponent with the EMP to make sure that unexpected impacts or failed mitigation measures are identified.

3.3. Management Strategies for Sustainable Development

The perception of sustainable development as applied today dates back to the early 1980s. Of all the definitions offered since then, the one expressed by the United Nations World Commission on Environment and Development (WCED) in its report *Our Common Future* still looks to be the most broadly known and recognized: "development that meets the needs of the present without compromising the ability of future generations to meet their own needs."[57]

Sustainable development provides a vision of progress that integrates immediate and longer term objectives, local and global action, and regards social, economic, and environmental issues as inextricable and interrelated components of systems' progress. In 2007, Eurostat also published a monitoring report on the basis of an extended set of sustainable development indicators. This report was also taken into account as one of the inputs for the first progress report on the sustainable development strategy in July 2009.[57, 58] For the enterprises and manufacturing systems, *sustainable development* refers to adopting business and manufacturing systems' strategies and activities that satisfy the needs of the enterprise and its stakeholders today, whereas supporting, sustaining, and boosting the human and natural resources that will be required in the future. Such a definition takes the spirit of the concept as originally presented by the WCED and identifies that economic development must satisfy the needs of a business enterprise/manufacturing system and its stakeholders. The latter includes shareholders, lenders, customers, employees, suppliers, and communities affected by the system's activities. It also emphasizes the business/manufacturing system's reliance on human and natural resources as well as physical and financial capital. It highlights that economic activity should not irreparably degrade or destroy these natural and human resources.

Moreover, it is essential to underline that sustainable development cannot be achieved by a single enterprise (or, for that matter, by the entire business community) in isolation. Sustainable development is an all-encompassing philosophy to which every participant in the global economy (containing consumers and government) must subscribe, if we are to meet today's requirements without compromising the ability of future generations to satisfy their own.

Recently, increased attention has been given to the role of organizations and manufacturing systems in implementing sustainable development. Sustainable operations need tactics and frameworks that can advocate organizational and industrial initiatives, whereas meeting multiple objectives such as maximizing financial performance and shareholder values while eliminating or minimizing potentially negative environmental externalities. Management strategies for sustainable development can be categorized as goal-oriented, objective-oriented, long-term (continuous), and short-term strategies.

3.3.1. Key Principles of Sustainable Development Strategies

Sustainable development strategies need systematic methodologies and iterative processes of learning by doing. Key principles of sustainable development strategies can be listed as follows:

1. Consensus on long-term vision: Tactical planning frameworks are more prone to be successful when they are on the basis of a long-term vision with a clear time frame in accordance with the agreement between a given manufacturer and his or her stakeholders. Simultaneously, in order to create a more accurate tactical planning framework, it is necessary to include effective policies and innovative ways to take short- and medium-term needs and changes into consideration as well.
2. Comprehensive and integrated: Strategies should pursue to integrate, where possible, economic, social, and environmental objectives. But where integration cannot be achieved, trade-offs require to be negotiated. Targeted with clear budget priorities, a sustainable development strategy should be fully integrated with existing budget processes to make sure that plans have the financial resources to attain their targets and do not represent mere "wish lists." On the contrary, the formulation of budgets should be informed by a clear identification of priorities. Capacity constraints and time limitations will have an impact on the extent to which the intended results are achieved. Objectives should be realistic concerning the constraints. A strategy should be established and advanced on existing capacity.
3. In accordance with a comprehensive and trustworthy assessment, recognition of priorities should be on the basis of a comprehensive analysis of the present situation, predicted trends and risks, and the links between local, national, and global challenges. External pressures on a manufacturing system, such as those resulting from globalization, or the impacts of global climate change should be factored into this analysis. Such analysis needs credible and reliable information on modifying environmental, social, and economic conditions, pressures, and responses, and their correlations with strategic objectives and indicators. Capacities for analysis and existing information should be fully applied, and various perceptions among stakeholders should be reflected.
4. Include monitoring, learning, and continuous improvement: Monitoring and evaluation require being in accordance with evident indicators and built into strategies to steer processes, track progress, distill and capture lessons, and signal when a modification on direction is necessary.

5. Building on existing processes and strategies (if it is possible): A strategy for sustainable development can be developed on strategies that previously exist in the manufacturing system, thus enabling convergence and consistency between different planning frameworks and policies. This requires efficient management to make sure of coordination of mechanisms and processes and to recognize and resolve potential conflicts.
6. Effective participation: Broad participation assists to open up debate to new thoughts and sources of information; expose issues that require to be addressed; enable problems, needs, and preferences to be expressed; recognize the capabilities required to address them; and develop a consensus on the need for action that leads to better implementation.

The objective of such strategies for manufacturing systems is to establish a balance between the three key factors, or three pillars: economic benefits, technological progress, and environmental protection accompanied by reasonable use of natural resources. Proactive environmental management strategies and initiatives, such as pollution prevention (P2), cleaner production (CP) for sustainable development, are inherently multi-objective processes that require joint considerations of environmental, industrial, economic, and social criteria in all stages of decision making. The success of such strategies and initiatives, though, depends on the strength and the relevance of their strategic frameworks.

3.3.2. Environmental Management System Development and Sustainable Environmental Management System Development

In this section, we deal with the similarities and differences of design steps for the environmental management system (EMS) development and sustainable environmental management system (SEMS) development frameworks.[59]

Step 1: Setting Goals and Objectives. The objective of this step is to develop a set of measurable goals, objectives, and criteria that can lead a manufacturing system toward the ultimate goal of reducing its negative externalities. This first step is crucial, as it provides direction, focus, and priorities.

- **EMS:** Characterizing the waste stream, setting waste reduction goals and objectives, and reducing negative environmental aspects.

- **SEMS:** Characterizing the waste stream, setting waste reduction goals and objectives, selecting proper performance criteria, and reducing environmental aspects in acceptable economic and social manners.

Step 2: Developing a Policy. The objective of this step is to develop a policy that states the manufacturing system's commitment to addressing environmental aspects and impacts of operations.

- **EMS:** Developing an environmental policy that focuses on maximizing waste reduction and environmental benefits.
- **SEMS:** Developing a sustainability policy that focuses on maximizing cumulative environmental, economic, and social benefits.

Step 3: Designing the Team. The ideas within the policy may be generated by collecting input from a management team put together by the EMS or SEMS champion.

- **EMS:** EMS team consists of internal stakeholders, focusing on managing environmental aspects of operations.
- **SEMS:** SEMS team consists of internal and external stakeholders and decision makers. They focus on managing operations and processes with consideration of economic social, and environmental factors.

Step 4: Designing Processes/Operations and Audit Programs. The objective of this step is to comprehensively identify and document information on the industry and operation.

- **EMS:** Evaluating sources, types, characteristics, and environmental as pects/impacts of emissions.
- **SEMS:** Evaluating sources, types, characteristics, and environmental, economic, and social impacts of operations and emissions using a system approach.

Step 5: Collecting Data. The objective of this step is to gather detailed data, in context of the determined goals and objectives. The types of tools applied differ across the manufacturing systems in their platform, sophistication, and association with other related functions.

- **EMS:** Collecting data on environmental emissions (sources, types, and characteristics).

- **SEMS:** Collecting data for operations: using a "system approach" focusing on, material, energy, water use, and environmental emissions (wastes), in addition to operation/process efficiency and potential data on the economic/social impacts of the operations and wastes.

Step 6: Assessing Data: Aspect Analysis. Data analysis focuses on qualitative and quantitative assessment of the environmental aspects of operations.

- **EMS:** Developing and recording environmental aspects of operation (the base case scenario).
- **SEMS:** Developing and recording environmental, economic, and social aspects of the operation (the base case scenario).

Step 7: Assessing Data: Impact Analysis. Data analysis concentrates on qualitative and quantitative assessment of the environmental impacts of the operations.

- **EMS:** Evaluating environmental impacts of operations as is (base case scenario).
- **SEMS:** Evaluating environmental, economic, and social impacts of the operation as is (the base case scenario).

Step 8: Designing and Characterizing Environmental Management Solutions. Developing and documenting emission management strategies according to the development goals and objectives of waste minimization, as stated in the environmental policy. Economic factors are a significant component in the selection of a solution.

- **EMS:** Designing strategies for emission elimination/reduction via P2/CP, and/or controlling, and estimating their efficiency, effectiveness, benefits, and costs.
- **SEMS:** Designing strategies for emission elimination/reduction via P2/CP, and/or controlling, followed by a quantitative and qualitative assessment of their integrated impact on environmental, economic, and social dimensions. Using multi-criteria decision making methods to select the best strategies based on the performance criteria, indicators, weight and priority of strategies, and threshold impacts that are defined by the SEMS team.

Step 9: Selecting Solutions. The objective of this step is to recognize the most effective and economical solution.

- **EMS:** Manufacturing system should search for a number of potential waste management strategies and choose the one that is most cost-effective.
- **SEMS:** Manufacturing system should explore a number of potential waste management strategies, assessing and ranking those according to their cumulative impact defined by an environmentally/economically/socially weighted score.

Step 10: Implementing Solutions. The purpose of this step is to put in place the solutions by accomplishing projects selected in Steps 8–9.

- **EMS:** Implementing the most cost-effective strategy.
- **SEMS:** Implementing the most sustainable strategy defined via ranking scores.

Step 11: Evaluating Solutions. The purpose of this step is to evaluate if implementation of the chosen waste management strategies resulted in satisfying desired/stated environmental management goals and objectives described in the policy statement.

- **EMS:** Evaluating extent of environmental emission reductions.
- **SEMS:** Evaluating the cumulative impact of implemented strategy on the environment, economics, and social dimensions.

Step 12: Adjusting and Modifying Solution. The purpose of this step is to make adjustments or modifications to the implemented solutions based upon the evaluation conducted in step 11.

- **EMS:** Modifying/making changes as needed, such as switching to new technologies, changing the budget, and modifying the implementation mode.
- **SEMS:** Modifying/making changes as needed, such as switching to new technologies, changing the budget, and modifying implementation mode, criteria, and performance indicators.

3.3.3. Analytical Techniques and Practices as Useful Tools for Selecting Proper Strategies

Analytical techniques and practices such as SWOT (strengths, weaknesses, opportunities, and threats) analysis, cost-benefit analysis, and multi-criteria analysis can help manufacturing systems to combine environmental, economic, and social considerations in their strategies.

3.3.3.1. SWOT Analysis

SWOT analysis, which is a useful technique for understanding the strengths and weaknesses—and for identifying both the opportunities open to a manufacturing system and the threats the manufacturing system faces—can be applied for selecting proper managerial strategies regarding sustainable development. It makes it possible to recognize the positive and negative factors influencing the achievement of an objective regarding sustainable development. Users of SWOT analysis require asking and answering questions that generate meaningful information for each category (strengths, weaknesses, opportunities, and threats) to make the assessment efficient and find their competitive advantage. Various elements of SWOT analysis are demonstrated in Table 3.8.

Table 3.8. Various Elements of SWOT Analysis

SWOT Analysis	**Strengths** 1. 2. 3. 4. . .	**Weaknesses** 1. 2. 3. 4. . .
Opportunities 1. 2. 3. 4. . .	Opportunities-Strengths Strategies (Use strengths to take advantage of opportunities) 1. 2. . .	Opportunities-Weaknesses Strategies (Overcome weaknesses by taking advantage of opportunities) 1. 2. . .
Threats 1. 2. 3. 4. . .	Threats-Strengths Strategies (Use strengths to avoid threats) 1. 2. . . .	Threats-Weaknesses Strategies (Minimize weaknesses to avoid threats) 1. 2. . .

3.3.3.2. Sustainability SWOT Analysis

The sustainability SWOT (S-SWOT) analysis is planned to deal with environmental challenges creating real manufacturing system risks and opportunities. It assists individuals to involve and motivate colleagues, especially those with limited information on environmental issues or corporate sustainability. This directing tool can be used to work across internal departments, as well as suppliers, customers, or other stakeholders on strategies to produce and sustain long-term value.

The S-SWOT analysis encourages teams to take wider connections and opportunities into consideration for collaboration inside and outside the manufacturing system in terms of sustainable development. It challenges teams to ponder the long-term environmental issues (such as natural resource scarcity) that will result in significant industrial risks or opportunities in the years ahead. These changes could arguably fit into a conventional SWOT analysis: an assessment of a manufacturing system's internal strengths and weaknesses, in relation to the external opportunities and threats. But studies on conducting SWOT analyses with a special emphasis on sustainable development and life cycle analysis show that manufacturing systems can identify new risks and opportunities. S-SWOT can also concentrate on how environmental threats impact not only the given manufacturing system but also other important stakeholders, for example, customers and communities. These can represent risks as well as opportunities to create new solutions for the given manufacturing system and the environment.

Moreover, S-SWOT can emphasize strengths that a manufacturing system already has and those a manufacturing system can make by collaborating with others (companies, customers, suppliers, communities). Similarly, it focuses on weaknesses that a manufacturing system can address with other stakeholders (and in some cases, competitors). A detailed S-SWOT analysis should be carried out to identify the manufacturing system's main strengths and weaknesses as well as the specific threats and opportunities. An example of strengths, weaknesses, opportunities, and threats regarding selection of proper sustainable development strategies is listed as follows:

Strengths

- Relatively strong financial position with a steady increase in profits
- Ability to produce higher value-added products
- Possibility to penetrate new developing markets

- Good reputation and a possibility to improve that image
- Good relationships with local communities
- Possibility to offset CO_2 emissions through forestry management
- Bold emphasis on health and safety
- Well-established quality control procedures
- Relatively high employee interest in the environment and sustainability

Weaknesses

- Emphasis on short-term returns and lack of long-term vision (10 years and beyond)
- Emphasis on quantity (of production) rather than provision of value-added products and services
- Limited understanding of the corporate sustainability concept and potential benefits
- Relatively high energy consumption and CO_2 emissions
- Relatively low awareness of the current developments regarding climate change and carbon trading resulting in a loss of income and a potentially negative image
- No formal environmental policy
- Lack of formalized environmental management systems
- The company car policy does not encourage the use of more sustainable transport modes

Opportunities

- Building on the existing image and taking the lead in sustainability in the group and/or in the sector
- Further improvement of the relationship with local communities through commitment to sustainability
- Reducing environmental risks and incidents and future liabilities
- Improving relationships with investors and customers through sound environmental, social, and ethical record leading to financial benefits

Threats

- Increased public awareness of sustainability and pressure-group activities
- Continued lack of understanding of key sustainability issues and areas of business that have an impact on sustainability and inability to respond to government and other initiatives on sustainability
- Further loss of income owing to increasing costs of "green" taxes
- Lack of own data on environmental and social performance, making the company more open to criticism and less able to refute various claims
- Loss of customers owing to the adoption of the sustainability principles further up in the supply chain and inability of the company to respond adequately
- Inability to penetrate new markets in the longer term through poor environmental and social image

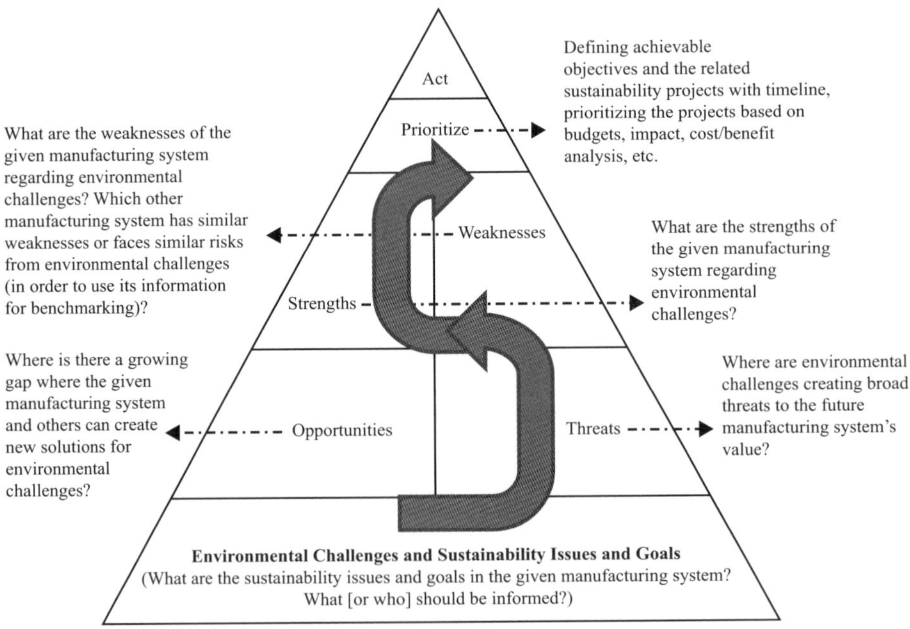

Figure 3.22. Various Elements of Sustainability SWOT (S-SWOT) Analysis

Moreover, various elements of S-SWOT analysis are shown in Figure 3.22. A sample of the S-SWOT analysis is also demonstrated through Table 3.9.

Manufacturing systems that "road-tested" the S-SWOT note that it creates a "common language" for involving colleagues and a "bridge" for thinking about a manufacturing system's present operation, its current sustainable programs, and where it needs to be in order to be competitive in future markets. The World Resources Institute (WRI), which is a global research organization that works closely with leaders to turn big ideas into action to sustain a healthy environment, reports successful cases that have applied S-SWOT analysis periodically. For instance, based on research conducted in December 2012, Delphi Automotive—a leading global supplier

Table 3.9. A Sample of S-SWOT Analysis

S-SWOT Analysis	**Strength (*S*)** • More budget for investing in sustainable development (1st *S*) • Possibility of installing wind turbines based on conducting a cost-benefit analysis (2nd *S*) • Collecting cardboards and installing new balers (3rd *S*)	**Weakness (*W*)** • Large amounts of solid waste, such as paper, metal, and plastics, that can be recycled are sent to landfill (1st *W*)
Opportunity (*O*) • Identifying new local recycling centers that are established near the facility (1st *O*)	▪ Contacting local recycling centers regarding the baled cardboards and making profit by selling them (considering 3rd *S* and 1st *O*)	▪ Contacting local recycling centers and selling recyclable materials (considering 1st *W* and 1st *O*)
Threat (*T*) • Cost pressures on raw materials, energy resources (1st *T*) • Regulatory pressure for eliminating chemicals, water pollution, health risks (dealing with waste management and removing toxics), and reduction of greenhouse gases (2nd *T*)	▪ Developing alternatives for scarce natural resources (energy, materials), for instance installing a wind turbine and using wind energy (considering 2nd *S* and 1st *T*) ▪ Using a portion of assigned budget for leveraging real-time monitoring and communication tools to track pollution and eliminate chemicals (considering 1st *S* and 2nd *T*)	▪ By selling recyclable materials, the manufacturing system can probably overcome cost pressures on raw materials and energy sources to some extent (considering 1st *W* and 1st *T*) ▪ By sending recyclable materials to recycling centers, the manufacturing system can reduce air pollution and water pollution (as they avoid sending the recyclable materials to landfills) (considering 1st *W* and 2nd *T*)

of technologies for the automotive and commercial vehicle markets and headquartered in Gillingham, England—while looking for ways to achieve "zero waste to landfill" ambitions, focused on its strengths and applied S-SWOT analysis. Delphi's two major strengths were categorized as follows:

1. Information systems for waste management: A database with information on waste volumes, recycling metrics, and other essential indicators to trace performance.
2. Strong customer connections: Close working relationships with major auto manufacturers (key customers) and mutual interests in establishing waste management systems that will result in the reduction of operational costs.

Delphi identified that its information systems could enable a more comprehensive waste management approach, comparing practices at related facilities and identifying cost-effective new approaches. The company is exploring pilot projects to drive toward "zero waste to landfill" at its headquarters and at facilities across each of its regions.

The Environmental, Health & Safety team will be advancing a toolbox and using available metrics to underline the cost-saving potential. The strengths also pointed to a tactical opportunity to work with customers. By involving such companies as General Motors (GM) and Ford, Delphi can combine expertise on innovative waste management practices. For instance, Delphi may be able to work with a partner like GM to recognize recycling vendors in growth regions such as Asia Pacific or even to co-invest in waste management systems for production facilities located nearby one another. The WRI also reports that some outstanding companies such as DuPont, Siemens, and GE are now positioning themselves to offer efficient solutions toward various environmental challenges, such as finding effective ways to control climate change, useful techniques for sustainable food production, and suitable methods for broader access to clean energy.[60]

3.3.3.3. Cost-Benefit Analysis

As we mentioned in previous sections, cost-benefit analysis is an effective tool for planning and decision making for environmental management and sustainable development, although its clients have been criticized for using discount rates that place a lower value on the environmental benefits and costs to future generations than on benefits and costs to the current generation. Discount rates can make current environmental protection measures appear too costly in comparison with the future benefit of a healthy environment. However, managers using cost-benefit analysis can select a lower discount rate to better balance current and future benefits and costs. For instance, *The Economics of Climate Change,* a review prepared for the British government in 2006, applied the low discount rate of 0.1% per annum.[61]

3.3.3.4. Multi-Criteria Analysis

As mentioned earlier, multi-criteria analysis (MCA) and multi-criteria decision making (MCDM) techniques have also gained popularity in sustainable development for energy planning and solid waste management. These methods offer solutions to the problems involving conflicting and multiple objectives. Various methods on the basis of weighted averages, priority setting, outranking, fuzzy principles, and their combinations are applied for energy planning and solid waste recycling decisions. Conventional single-criteria decision making is normally aimed at the maximization of benefits with minimization of costs. These techniques offer better understanding of inherent characteristics of a decision problem, encourage the role of participants in decision-making processes, facilitate compromise and collective decisions, and provide an appropriate platform for understanding the perception of models and analysts in a realistic scenario. The techniques assist to enhance quality of decisions by making them more explicit, rational, and efficient. Negotiating, quantifying, and communicating the priorities are also facilitated by applying these methods. The need to include environmental and social considerations in sustainable development in energy planning and solid waste recycling programs is conducive to increased use of multi-criteria approaches.

MCDM methods are divided into multi-objective decision making and multi-attribute decision making. There are various techniques in each of these categories. Priority based, outranking, distance based, and mixed methods are also used for several problems.

Each method has its own features, and the methods can also be categorized as deterministic, stochastic, and fuzzy methods. Decision making under uncertainty and decision support system are also prominent decision-making methods. The commonly used MCDM methods in the sustainable development analysis are AHP (analytical hierarchy process), PROMETHEE (preference ranking organization method for enrichment evaluation), ELECTRE (the elimination and choice translating reality), MAUT (multi-attribute utility theory), ANP (analytic network process),

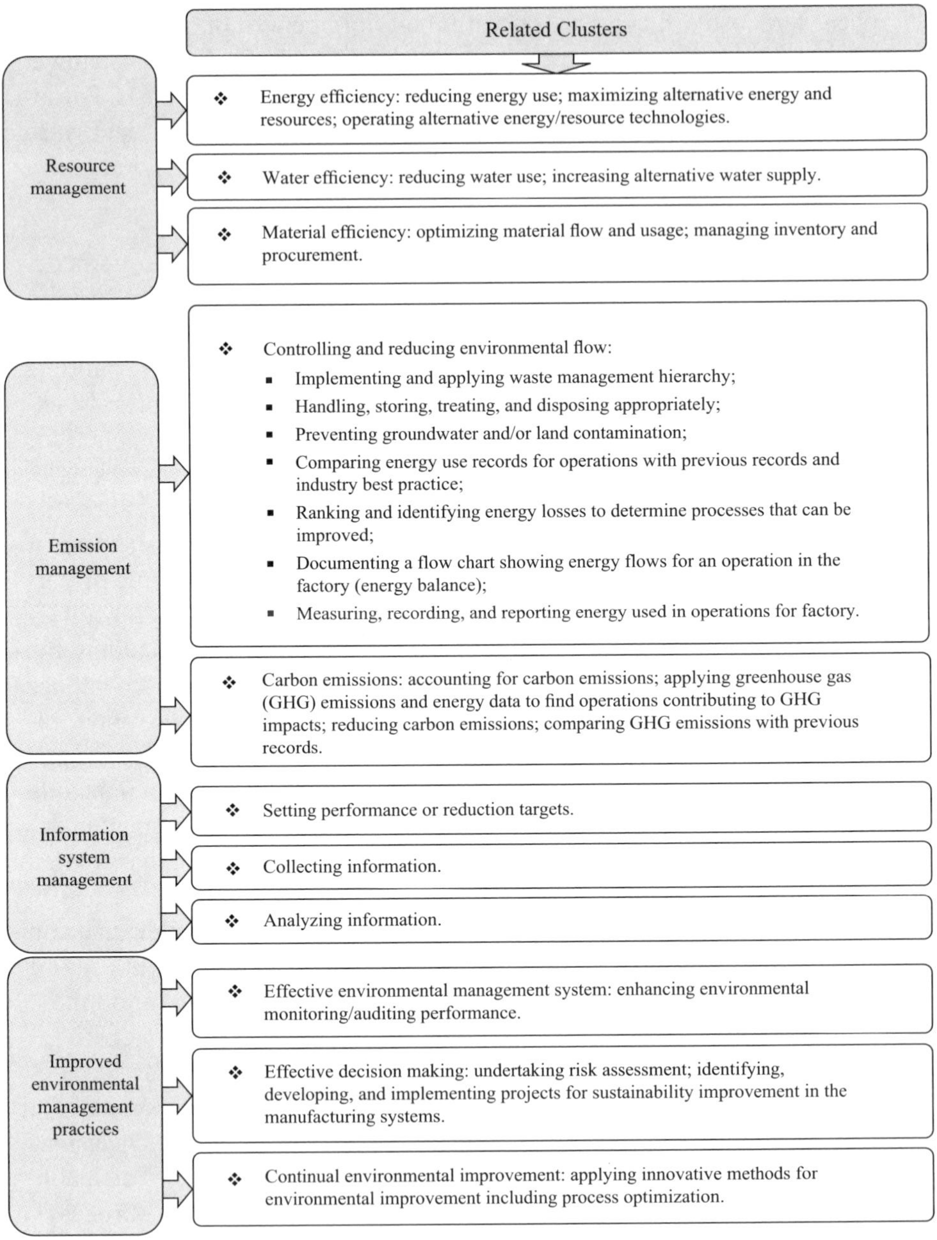

Figure 3.23. Management Framework for Sustainable Development in a Manufacturing System

CP (compromise programming), and TOPSIS (technique for order preference by similarity to ideal solutions).[62]

3.3.4. Management Framework for Sustainable Development

The management framework for sustainable development offers some practical guidance to manufacturing, corporate, and service levels in delivering sustainable development. It attempts to make sustainable development relevant and simple and to illustrate that its principles and outcomes lie at the heart of efficient governance. An all-inclusive management framework for sustainable development in a manufacturing system is presented in Figure 3.23.

This framework is designed to particularly target the key elements of resource management, emissions management, information system management, and environmental management to assist manufacturers in reducing their environmental footprint and meeting their strategic measurable

Table 3.10. A Sample Checklist for Sustainable Development

Sustainability metrics	Questions
a) Material use	Are the structure materials made from recycled or reclaimed material? Are the structure materials renewable resources?
b) Energy use	How much energy is used in materials' production process? Are the materials locally available, minimizing energy needed for transport?
c) Water use	How much water is used in materials' production process?
d) Solid waste generation	Does material production (or acquisition) produce a low amount of solid waste?
e) Emissions generation	Does material production produce low traditional air pollutant emissions (particulates, sulfur dioxide, nitrogen oxides, etc.)? Does material production produce low greenhouse gas emissions (carbon dioxide)?
f) Toxic releases	Is the material nontoxic, which would generally lower the potential for toxic releases?
g) Land impact	Does material production (or acquisition) and transport produce minimal impact on plants, animals, and ecosystems?
Other	Specify: ....

environmental targets. It can also be applied as a platform to advance the capability metrics.

Furthermore, in order to make a better management framework for sustainable development of a manufacturing system, a sustainable development checklist can also be designed. It can be considered as a method for comparing alternatives on the basis of project/product life cycles in a manufacturing system. The checklist includes various tables, one for each life cycle stage and one summary table. Each of the tables can list questions about material use, land impacts, emissions, and wastes for each life cycle stage. A sample checklist for sustainable development is shown in Table 3.10.

Figure 3.24 also describes the approach for sustainable development assessment of a production work cell. The inputs of the work cell include physical flows and information flows. Process conditions (information flows) include process settings (e.g., machining parameters) and working environment parameters (e.g., lighting and heating levels). The outputs of the work cell include the processed part, solid and liquid waste, which may be associated with economic, environmental, and social impacts. The ver-

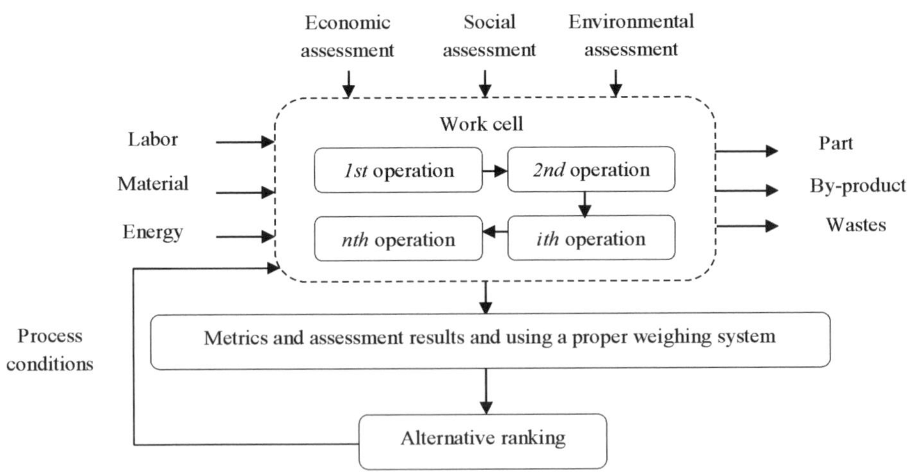

Figure 3.24. Approach for Sustainable Development Assessment of a Production Work Cell

tical flows in the figure depict the information flows where assessment tools and decision making are involved. Economic, environmental, and social impacts are analyzed. The results are integrated using an alternative ranking MCDM process and improved process settings can be identified and chosen as a new working condition.

The economic assessment covers four work cell aspects: facility costs, labor costs, material costs, and utility costs. In order to compare options, decision makers are required to identify and quantify appropriate metrics. Economic metrics are computed for a cost per part basis. Cost per part can represent a factory's economic goal, which is to gain maximum profit based on limited orders. Economic metrics can also be computed for a cost per time period. This happens when the work of a work cell is not affected by the lack of product orders (i.e., it is fully utilized). The outcome of cost assessment can be a single value (e.g., in dollars). It can also be multiple values in different categories, which gives more detailed information for cost analysis of specific materials.

Life cycle assessment (LCA) can be used to analyze the environmental impacts of the work cell. By identifying the quantity of materials and energy use for different processes in the work cell based upon a defined functional unit, practitioners can associate the input and output materials and energy use with environmental impact categories for scenario-based comparisons.

Life cycle inventory (LCI) analysis also comprises creating an inventory of flows within the work cell, including raw material, energy, releases, and any resources related to the work cell processes. Life cycle impact assessment (LCIA) evaluates the significance of environmental impacts identified in the life cycle inventory. In this phase, the LCI outcomes are typified to produce numerous impact indicators.

Assessment of social sustainability at the work cell level has more challenges than economic and environmental sustainability. For social impact assessment, stakeholders can be categorized into the following types: value chain actors, society, worker, consumer, and local community.

To evaluate sustainable development of products in a manufacturing system and have a clear view regarding that, related sustainability scores can be defined to be able to compare and select between different alternative designs, or between different production scenarios. One simplified yet comprehensive and generalized procedure for such a computation is to calculate the Product Sustainability Index (PSI) by computing the weighted sum of different subelements within the three main sustainable development element categories (environmental, societal, and economic impacts) for each of the four life cycle stages: premanufacturing, manufacturing, use, and post-use. The product designers document the scores of the influencing factors in each box of the matrix and evaluate the PSI by proceeding across the matrix and summing the scores of all influencing factors in each matrix category to calculate the percentage value. For instance, the PSI for the environment component of the premanufacturing stage ($PSI_{(en\_pm)}$) can be computed as the following equation:

$$PSI_{en\_pm} = \left\{ \frac{\left[\sum_{i=1}^{n} IF_{en\_pm} \right]}{n \times 10} \right\} \times 100\% \tag{33}$$

Where $IF_{(en\_pm)}$ refers to the influencing factor rated on a scale of 0–10 for the environment component of the premanufacturing stage and n refers to the number of influencing factors. The PSI values for society ($PSI_{(so\_pm)}$) the economy ($PSI_{(ec\_pm)}$), which are components of the premanufacturing stage, can be calculated similarly. The PSI values for any life cycle stage, for instance, premanufacturing, can be evaluated by averaging the PSIs of sustainability components in the vertical column for that particular life cycle stage. Similarly, the PSI values of environment (PSI_{en}), society (PSI_{so}), or economy (PSI_{ec}) sustainability components across all four life cycle stages can be analyzed by averaging the PSIs of life cycle stages in the horizontal row for that specific sustainability component. The overall product sustainability index for a product over its total life cycle (PSI_{TLC})can then be computed as:

$$PSI_{TLC} = \frac{\left[PSI_{en} + PSI_{so} + PSI_{ec} \right]}{3} \tag{34}$$

According to the introduced index, a sample framework for a comprehensive product sustainability development evaluation is shown in Table 3.11 Please consider the numbers are based on an assumption and the integrated PSI in the last column and row, which is denoted by PSI_{TLC}, shows the computed total sustainability index with equal weighting for all ele-

Table 3.11. A Sample Framework for a Comprehensive Product Sustainability Development Evaluation

Life cycle stages		Premanufacturing (score out of 10)		Manufacturing (score out of 10)		Use (score out of 10)		Post-use (score out of 10)		Total scores
Sustainable development components	Environment	Material extraction	7	Production energy used	7	Emissions	9	Recyclability	7	$PSI_{(en)}$ 77.29
		Design for environment	8	Hazardous waste produced	9	Functionality	8	Re-manufacturability	8	
		Material processing	6	Renewable energy used	8	Hazardous waste generated	9	Redesign	7	
								Landfill contribution	7	
		$PSI_{(en\_pm)}$	70	$PSI_{(en\_m)}$	80	$PSI_{(en\_u)}$	86.67	$PSI_{(en\_pu)}$	72.5	
	Society	Worker health	8	Work ethics	7	Product pricing	7	Take-back options	7	$PSI_{(so)}$ 73.54
		Work safety	8	Ergonomics	7	Human safety	9	Reuse	6	
		Ergonomics	7	Work safety	8	Upgradeability	7	Recovery	7	
		$PSI_{(so\_pm)}$	76.67	$PSI_{(so\_m)}$	73.33	Complaints	8	$PSI_{(so\_pu)}$	66.67	
						$PSI_{(so\_u)}$	77.5			
	Economy	Raw material cost	6	Production cost	6	Maintenance cost	7	Recycling cost	7	$PSI_{(ec)}$ 61.25
		Labor cost	3	Packaging cost	7	Repair cost	6	Disassembly cost	8	
		$PSI_{(ec\_pm)}$	45	Energy cost	8	Consumer injury cost	8	Disposal cost	4	
				Transportation cost	6	Consumer warranty cost	7	Remanufacturing cost	7	
				$PSI_{(ec\_m)}$	65	$PSI_{(ec\_u)}$	70	$PSI_{(ec\_pu)}$	65	
Total scores		$PSI_{(pm)}$	63.89	PSI_m	72.78	$PSI_{(u)}$	78.06	$PSI_{(pu)}$	68.06	$PSI_{(TLC)}$ 70.69

ments and subelements. If a sustainability analyst in a given manufacturing system prefers to prioritize sustainability elements based on the unique needs for a particular design scenario, he or she can apply a proper MCDM method to consider deterministic or fuzzy weighing systems for elements and subelements.[63, 64]

3.3.5. Engineering Economics and Project Prioritization

Engineering economics analysis is a critical component of a comprehensive project or program evaluation methodology that can consider the majority of key quantitative and qualitative impacts of sustainability investments. The basic concept and methodology of economics analysis were developed at the beginning of the 20th century and were applied in different industries. Since that time, several static and dynamic models have been proposed and are currently available to help in economics analyses. Static models can predict short-term improvements. Typically, a static model is utilized to take a snapshot in time of how an economic effect can ripple throughout the economy of a region. Static models do not have the power of future (dynamic) prediction of the economy of an area. Dynamic models, in contrast, can simulate the effects of several factors in the economy and forecast how those factors will change the economy over time.[65]

Economics analysis is now commonly applied in the evaluation of the cost efficiency of various types of projects, including highway and bridge projects, construction projects, manufacturing projects, facility-location projects, and sustainability projects. It allows analysts to recognize, quantify, and value the economic benefits and costs of defined projects and programs over a multiyear time frame. By applying this information, project planners are able to target scarce resources to their best uses in terms of maximizing benefits to the public. It is vital in the sustainable development process that each alternative for sustainability and project is appropriately evaluated for its costs and benefits throughout its entire life cycle. Project planners utilize measures such as the net present value of costs and benefits, benefit-cost ratio, and the internal rate of return to compare various competing alternatives. The alternative that gives the highest net present value, benefit-cost ratio, or return on investment is chosen and considered to be funded, programmed, and eventually implemented.

The results of economics analysis enable sustainability managers and project planners to systematically and rationally prioritize sustainable development projects in terms of expected benefits and costs. Consequently, the limited funds and resources can be allocated in such a manner that maximizes benefits. Life cycle cost analysis (LCCA) and benefit-cost analysis (BCA), as we mentioned in previous sections, are also the two most commonly applied economics analysis methods for determining whether the proposed projects are worth undertaking. LCCA is used to evaluate the total cost differences within a facility's life cycle among several proposed alternative projects. BCA considers the costs and also evaluates the benefits incurred in the future. Thus, LCCA usually can be considered a cost subset of BCA.[66]

Sustainable project decisions need more inclusive prediction of future costs and benefits. These elements are exposed to uncertainty. In order to evaluate the sustainability and financial return on investment and demonstrate the functional, economic, social, and environmental value of a potential project for sustainable development, engineering economics analysis can be applied. Sustainable return on investment considers noncash benefits and costs of environmental and societal factors, such as lower emissions, that impact both the operation and the community as a whole (Fig. 3.25). Engineering economics analysis assigns monetary values to stimulus priority requirements to highlight the best and highest value projects. For instance, consider that a sustainability manager in a given manufacturing

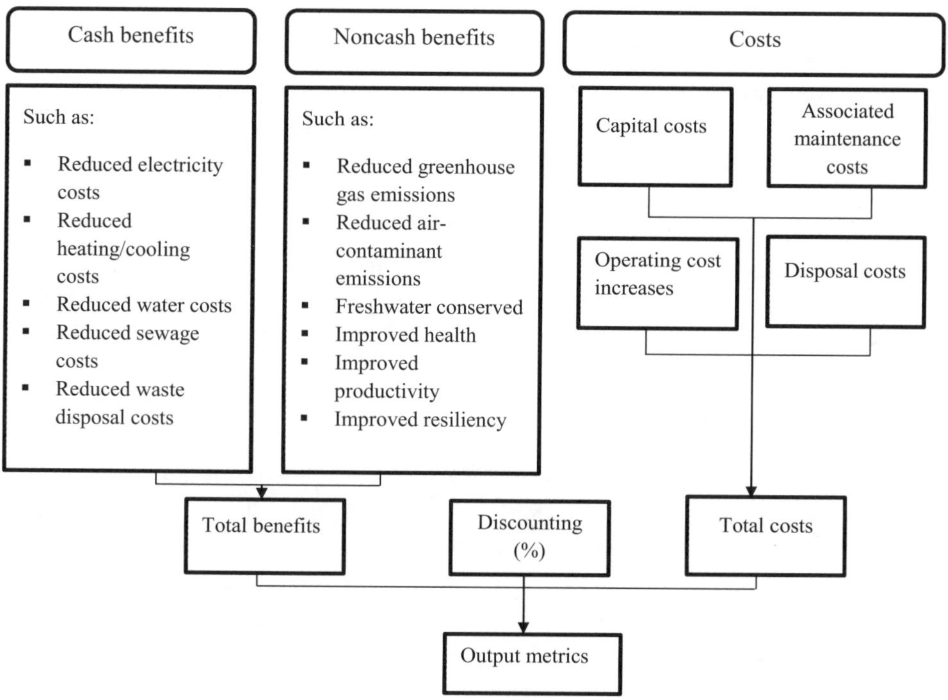

Figure 3.25. The Importance of Considering Sustainability and Financial Return in Evaluation of Potential Projects for Sustainable Development

system decides to start a project for sorting solid wastes and selling recyclable materials to potential recycling centers and companies. This individual can apply engineering economics analysis to do a complete evaluation regarding the initial investment in sorting solid waste and separation of recycling materials, related transportation costs for transporting materials to nearest the recycling centers, payback period, and achieved income by selling recycled materials. Some recent examples that have applied engineering economics regarding their sustainable development projects include Enbridge Gas Distribution, Toronto, Ontario (analysis of carbon emission trading); King County, Seattle, Washington (assessment of costs, benefits, and cost-sharing opportunities through water source exchange projects in Puget Sound); California Bay Delta, Northern California (assessing the costs and charting a transparent path to conserve, restore, and protect the diverse ecosystem and critical water supply); and Denver's Metro wastewater reclamation district, Denver, Colorado (using engineering economics to quantify the value of water and energy consumption, compute greenhouse gas and air-contaminant emissions, and evaluate the cost/benefits of using biosolids for fuel and the impact of odor on residents).

Before a sustainable development project receives funding for implementation, a project selection process must be followed. The list of potential projects should be prepared and accompanied by supporting information and a feasibility study, including location, cost, benefits, size of the project, existing infrastructure, and sustainability background in the given manufacturing system regarding the defined project. Once the potential project list is compiled, the scoring process can begin to rank the potential projects. Scoring from the databases should be completed as soon as all databases have been updated. Once the databases are compiled and updated, the scoring can be completed relatively quickly. Each project can be scored on its own, according to the defined criteria such as the result of engineering economics analysis. A project score model has been presented in Figure 3.26. After this stage, projects will be compared based on the gained scores. If there is a need to compare a set of projects, a mean of the economic scores should be considered.

Moreover, if the manufacturing system prefers to invest in more than one project and work on various projects for sustainable development at the same time, it can apply the project and portfolio management (PPM) procedure. A competitive advantage for manufacturing systems regarding

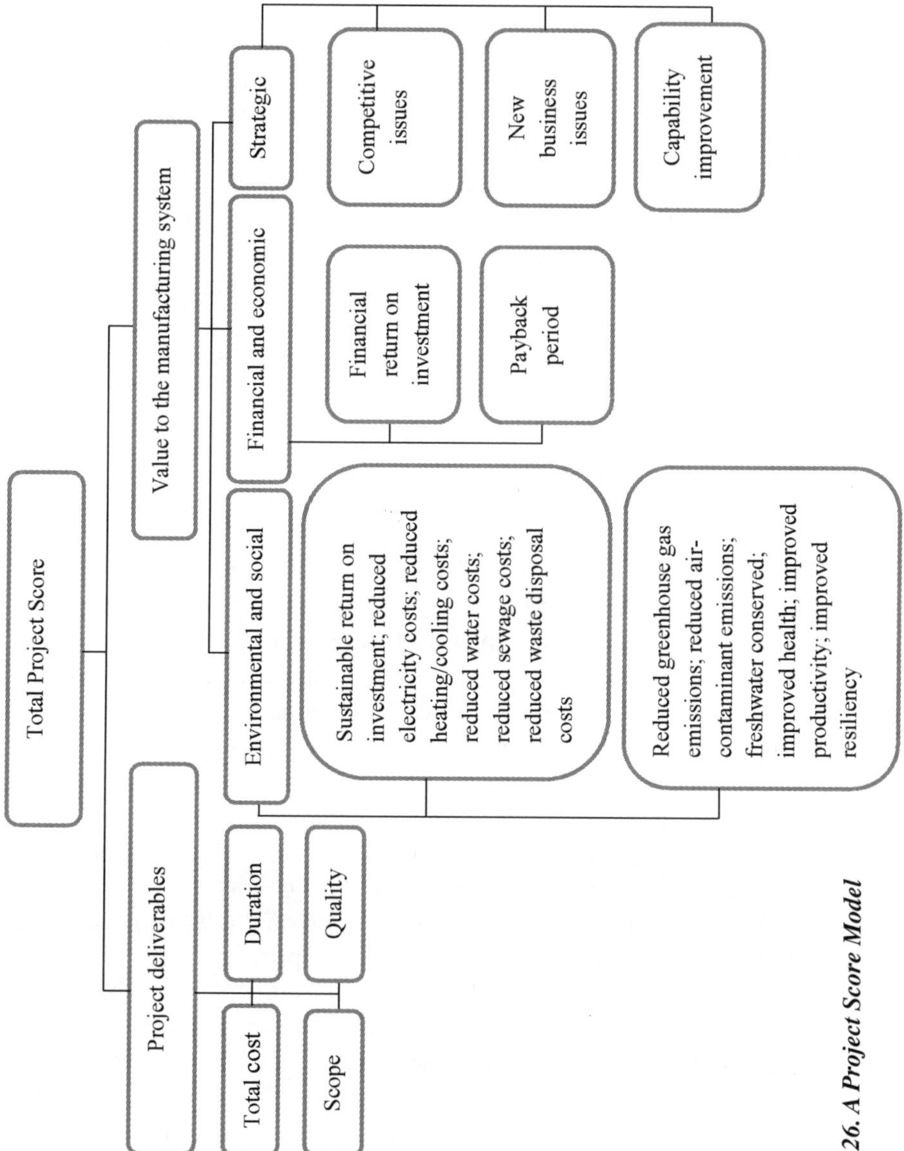

Figure 3.26. A Project Score Model

sustainable development projects is doing the right projects and making sure that there are resources to complete those projects. Selecting and allocating resources for the proper projects are essential factors in optimizing project outcome. A portfolio contributes to a collection of projects or programs that are grouped together to facilitate efficient management of that work to satisfy the targets of the tactical enterprise and manufacturing system. The projects or programs of the portfolio might not essentially be interrelated or directly linked. As a sample, an infrastructure firm that has the tactical objective of "optimizing the return on its investments" may provide a portfolio that entails a mix of projects in oil and gas, power, water, roads, rail, and airports. From this combination, the firm may select to handle the linked projects as one program. All of the power projects may be grouped together as a program for power. Likewise, all of the water projects may be grouped together as a water program. Portfolio management emphasizes making sure of this fact that projects and programs are reviewed to prioritize resource allocation and that the management of the portfolio is coherent with and aligned to the tactics of a manufacturing system or an enterprise.

PPM is a process and set of practices that enable manufacturing systems to handle projects as a strategic portfolio, ensuring the alignment of programs and projects with their manufacturing goals. Executives require a frequent review of entire portfolios and programs, decide which projects are or are not necessary, see where investment should be done, prioritize projects, distribute resources properly, and keep track of improvements.[67] Figure 3.27 refers to core PPM process overview. Figure 3.28 also demonstrates the importance of sustainable development project selection and PPM as the hub and their connection to other essential evaluation elements.

There are various methods for assessing the economic consequences of sustainable project decisions, comprising those on the basis of the projected cash flows (e.g., net present value or net present worth [NPV or NPW] and internal rate of return [IRR]) as well as those originating from standard accounting methods (e.g., payback and return on investment [ROI]). For sustainable development projects with substantial initial investments or capital costs, a detailed profitability engineering economic evaluation is required. As we mentioned earlier in chapter 2, the three standard measures of profitability are payback period, IRR, and NPV. The re-

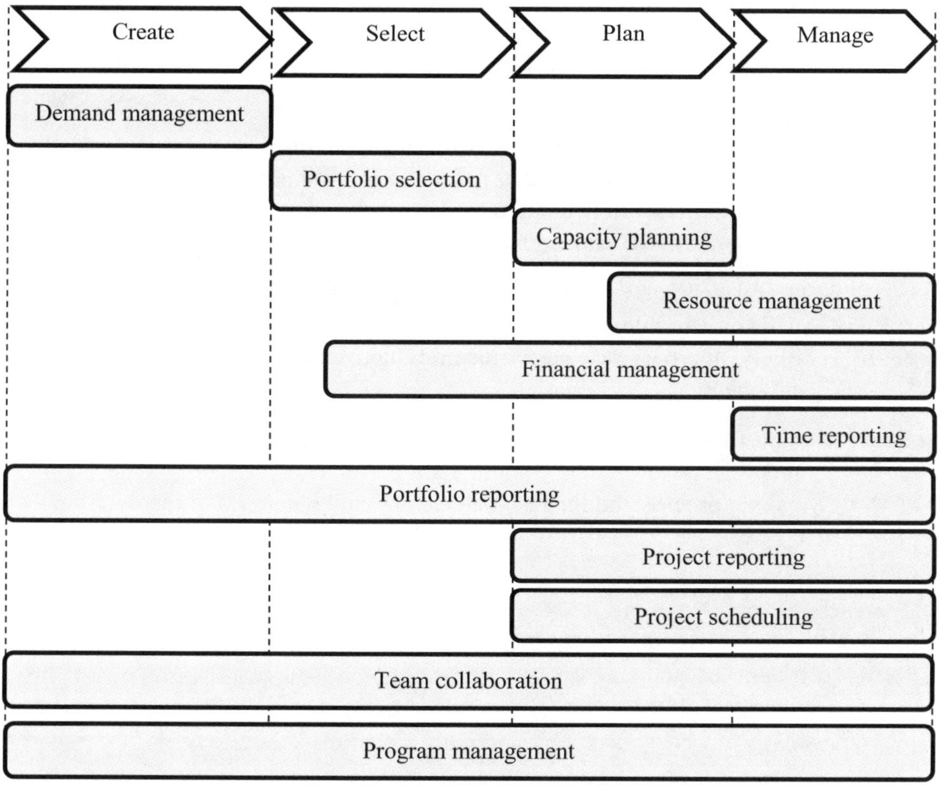

Figure 3.27. Core Project and Portfolio Management Processes Overview

lated equations are also mentioned in that chapter. In the next section, we present some examples for project investment appraisal.

3.3.5.1. Investment Appraisal Examples of Sustainable Development Projects

Investment appraisal refers to evaluations of decisions made by organizations and systems on allocating resources to investments of a significant size. Four key principles in investment appraisal of sustainable development projects can be categorized as follows:

1. When we are appraising multiperiod investments, we should estimate the expected benefits, costs, and related cash inflows and outflows that arise over time. In such a situation, the time value of money should be taken into account.

2. A satisfactory decision relies on an understanding of the system. Considering that, we need to gather sufficient information and perform a thorough interpretation in relation to the system's strategies and its economic, environmental, social, and competitive goals and objectives.

3. Useful techniques such as a sensitivity analysis should be applied to identify key variables and risks. Effective tools help to reflect worst, most likely, and best case scenarios and, therefore, can support a reasoned judgment.

4. A post-completion review or audit of an investment decision is required, and this review / audit should include an assessment of the decision-making process and the results, benefits, and outcomes of the decision.

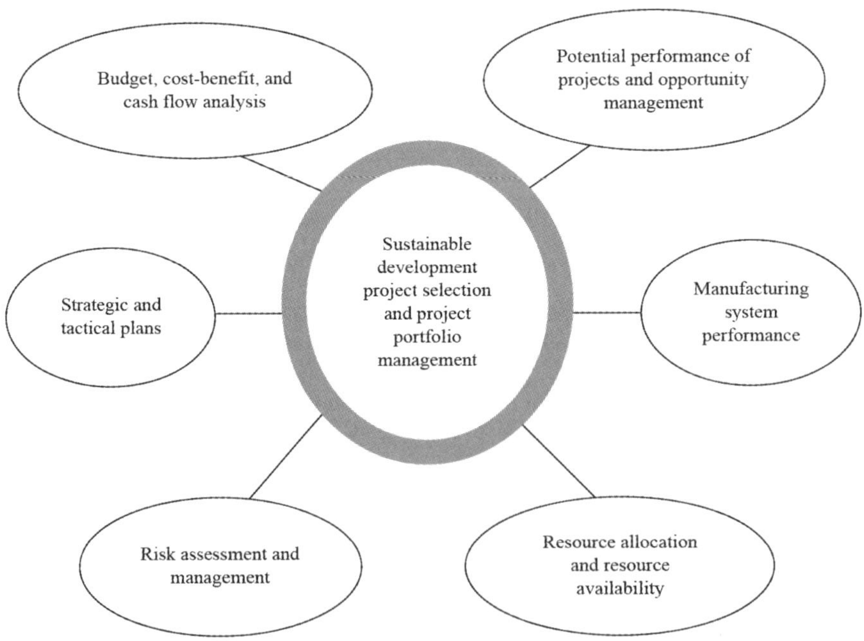

Figure 3.28. Connection of Sustainable Development Project Selection and Project Portfolio Management with Essential Evaluation Elements

Here, in order to somehow clarify the investment appraisal for sustainable development projects, we focus on five related examples.

First Example: Suppose that a potential sustainable project for a given manufacturing system is to install a cardboard baler for a total cost of $65,000. If the baler is expected to save the company $20,000 per year, then the payback period is (65,000/20,000) 3.25 years. The payback period is normally measured in years, but some alternatives may have payback periods based on months. Many enterprises apply the payback period as a screening technique earlier than performing a full financial evaluation. If the alternative does not satisfy a predetermined threshold, the alternative is rejected. Payback periods in the range of three to four years are usually considered acceptable for low risk investments. This method is recommended for quick assessments of profitability. If large capital expenditures are considered, a more strenuous financial evaluation such at the IRR and NPV is needed. IRR and NPV are both discounted cash flow methods for deciding profitability and determining if a waste minimization option will enhance the financial position of the company. Lots of organizations utilize these techniques for ranking capital projects that are competing for funds, such as the case with the different waste minimization options. Capital funding for a project can rely on the ability of the project to produce positive cash flows beyond the payback period to realize a satisfactory return on investment. Both the IRR and NPV identify the time value of money by discounting the anticipated future net cash flows to the present. For investments with a low level of risk, an after-tax IRR of 12–15% is normally acceptable. (The calculations of NPV and IRR are also described earlier in chapter 2.)

Most spreadsheet programs usually have the ability to automatically compute IRR and NPV from a series of cash flows. Regarding the baler installation project (with an initial cost of $65,000 and $20,000 in annual savings.), additionally, the assumed baler life span is 10 years and an organization minimum attractive rate of return (MARR) is 15%. The MARR is the minimum return on a project that a manager is willing to accept before starting a project, given its risk and the opportunity cost of foregoing other projects. The cash flows and IRR and NPV values are shown by Table 3.12.[68]

As shown in the last two rows of Table 3.12, the IRR is 28.2% and the NPV is nearly $31,000 at an assumed MARR of 15%. The facts that

Table 3.12. Computed Values for Cash Flow, NPV, and IRR in Baler Installation Project

Year	Cash flow
0	$(65,000)
1	$20,000
2	$20,000
3	$20,000
4	$20,000
5	$20,000
6	$20,000
7	$20,000
8	$20,000
9	$20,000
10	$20,000
IRR	28.2%
NPV	$30,761
MARR	15%

the IRR is greater than the 15% MARR and the NPV is positive indicate that the project is a good financial decision.

Second Example: Consider the example of a shop evaluating the purchase of a still to recycle its waste solvent. The shop manager evaluates both his current operation and the option of using a still. He notices that the installation of a still will cost $7,700 but provides a net annual operational savings of $4,634. Once the net annual savings is divided into the initial cost, the manager finds that the still will pay for itself in 1.7 years

($7,700/$4,634). Please note that in this example, it is assumed that the annual cash flow is the same each year. In reality, there are considerable costs such as depreciation and taxes that will result in the cash flow variation annually. If the annual cash flow differs from year to year, the payback period is determined when the accrued cash savings is equivalent to the initial investment costs (i.e., when the cumulative cash flow balance equals zero).

Third Example: Consider the two given sustainable development projects, A and B, in Table 3.13. To understand if Project A is a better investment opportunity than Project B, the manufacturer or solid waste manager should consider the following items:

(i) Cash flows: Project B has a higher total cash inflow ($24,000 vs. $15,000).

(ii) Total net profit: Project B has a higher total net profit over 3 years ($12,000 vs. $9,000).

(iii) Average annual profit (or return): Project B has a higher average annual return ($4,000 vs. $3,000).

(iv) Rate of return on investment: Project A has a higher rate of return on investment (50% vs. 33%).

Table 3.13. Annual Rate of Return, Project A vs. Project B

Evaluation factors	Project A	Project B
Investment, end of year 0	$(6,000)	$(12,000)
Cash flows, end of year 1	$3,000	$7,000
Cash flows, end of year 2	$4,000	$8,000
Cash flows, end of year 3	$8,000	$9,000
Total cash flows	$15,000	$24,000
Total net profit	$9,000	$12,000
Average annual profit	$3,000	$4,000
$AARR = \frac{average\ annual\ profit}{initial\ investment}$	$\$3{,}000/\$6{,}000 = 50\%$	$\$4{,}000/\$12{,}000 = 33\%$

The average annual rate of return on investment is simple, easy to understand, and a generally good project appraisal methodology. It is about the profitability of a capital project over the lifetime under consideration, which in this case is only 3 years. It does not give any information about the timing of cash streams that flow from the project. The averaging process removes such relevant information as timing; in this specific case, it does not consider that Project A has a rapidly increasing year-on-year cash inflow, while the returns from Project B are relatively static. Further, the investment needed for Project B is twice that needed for Project A.

Fourth Example: Consider the two given sustainable development projects C and D in Table 3.14, where the investment required for each sustainable development project is the same. In this example, the timing of the relative

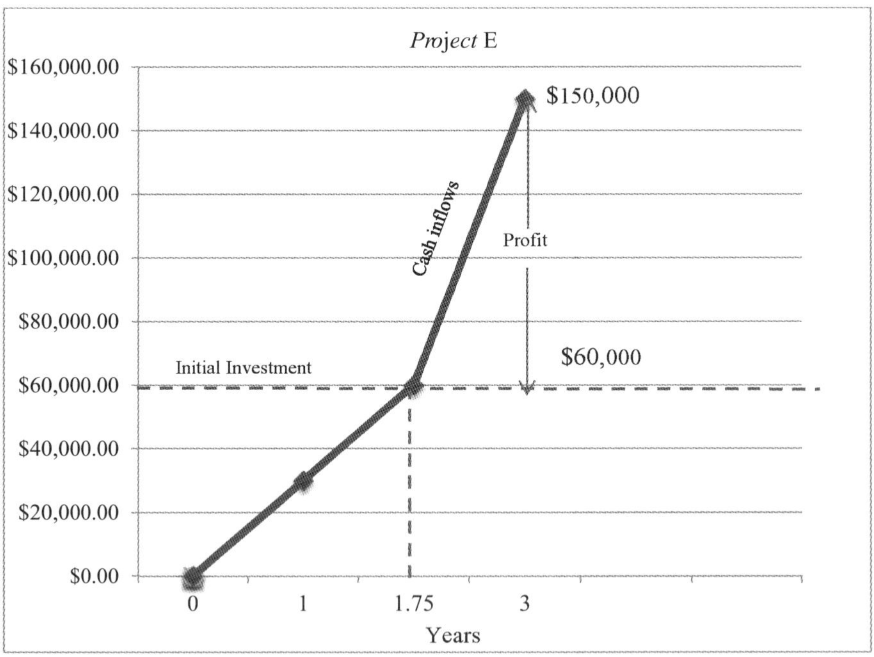

Figure 3.29. Payback Periods for Project E

Table 3.14. Annual Rate of Return, Project C vs. Project D

Evaluation factors	Project C	Project D
Investment, end of year 0	\$(60,000)	\$(60,000)
Cash flows, end of year 1	\$10,000	\$60,000
Cash flows, end of year 2	\$20,000	\$20,000
Cash flows, end of year 3	\$60,000	\$10,000
Total cash flows	\$90,000	\$90,000
Total net profit	\$30,000	\$30,000
Average annual profit	\$10,000	\$10,000
$AARR = \frac{average\ annual\ profit}{initial\ investment}$	$\$10{,}000/\$60{,}000 = 16.7\%$	$\$10{,}000/\$60{,}000 = 16.7\%$

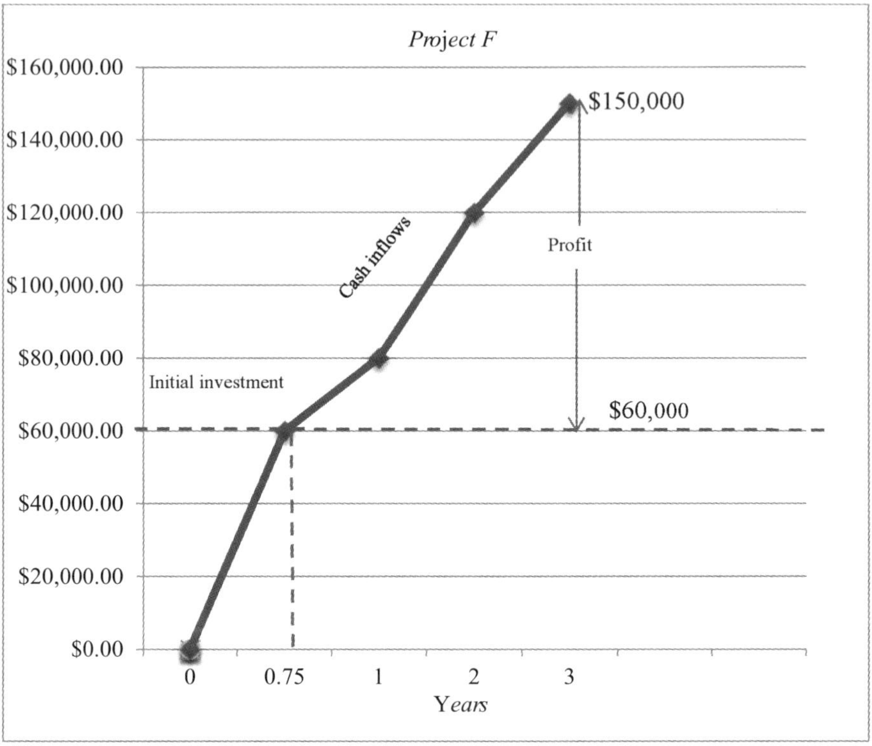

Figure 3.30. Payback Periods for Project F

magnitudes of the annual income streams is in reverse order but the total income over the 3-year project appraisal period is the same. The average annual rate of return on investment for each project is now also the same, although it is obviously sensible to get the return sooner than later. Under Project D, the extra $50,000 income received at the end of the first year could be taken out of the project and used fully invested for 2 years in the capital markets to yield a further positive return. Therefore, considering what is known as the opportunity cost of capital, Project D is superior.

Fifth Example: Consider the data of two potential sustainability projects, E and F, that are presented in Table 3.15. The payback period for Project E can be easily calculated. $30,000 is rapid in year 1, and $40,000 in year 2. Thus, Project E's payback period is years (1 year at $30,000 + three-quar-

Table 3.15. Cash Flows for Project E and Project F

Evaluation factors	Project E	Project F
Investment, end of year 0	$(60,000)	$(60,000)
Cash flows, end of year 1	$30,000	$80,000
Cash flows, end of year 2	$40,000	$40,000
Cash flows, end of year 3	$80,000	$3,000

ters of a year at $40,000 equals the initial investment of $60,000). Similarly, Project F's payback period can be calculated at 9 months. Only three-quarters of the $80,000 cash flow in year 1 is needed to recover the initial investment of $60,000 (assuming that cash is received evenly throughout the year). These results are shown graphically in Figure 3.29 and Figure 3.30.

The payback technique has one benefit over the average rate of return on investment: It does take time into account. Project F's payback period is 9 months and Project E's is 1 3/4 . However, one has not set a payback period target by which different projects are judged. In other words, one has not considered the maximum payback period acceptable or upon what criteria such judgements are based as being important. The payback method ignores cash receipts accepted after payback. This could be essential when comparing project viability, particularly if one project has a speedily improving cash stream over time. That project's payback method is, hence, a measure of risk but not of profitability. It may be considered a rough screening device for evaluating which projects to invest in.

3.3.5.2. Multiknapsack Model

When practitioners and managers plan for capital budgeting, they often form a priority list of candidate projects, by scoring the projects individually, using economic measures such as net present value, benefit investment ratio, payback period, and internal return rate. The multiknapsack methodology to capital budgeting takes as input a budget forecast, in company with the stream of liabilities and the profit of each project.[69]

The multiknapsack model is an integer programming that has a binary decision variable for each project to determine whether it is chosen or not;

a budget limitation for each time period (e.g., year); and the objective of maximizing total profit. The output of the multiknapsack model is a set of projects to be carried out, assuming the point forecasts for these input parameters are correct. This chosen set of projects is considered as a project portfolio. If the costs and profits of the candidate projects in addition to the budgets in coming years know with certainty, the multiknapsack model offers an attractive tool for choosing a project portfolio.

Set of Notation and Multiknapsack Formulation for the Deterministic Capital Budgeting Problem: The notation and formulation of the multiknapsack model are as follows:

Indices and Sets:

$i \in I$ Candidate projects

$t \in T$ Candidate periods (years)

Data:

a_i: Net present value of project i

b_t: Available budget in year t

$c_{i,t}$: Cost of project i in year t

Decision variables:

x_i: 1 if project i is chosen; 0 otherwise

Formulation:

Objective function:

$$\max_x \sum_{i \in I} a_i x_i \tag{35}$$

Subject to :

$$\sum_{i \in I} c_{i,t} x_i \le b_t \qquad t \in T \tag{36}$$

$$x_i \in \{0,1\} \qquad i \in I \tag{37}$$

The objective function sums up the NPV contributions of all chosen projects. The first constraint makes sure that the annual cost of the chosen projects is within the budget for each year b_t $t \in T$. Yes/no restrictions on selecting projects are enforced by the second constraint. The optimal solution to the multiknapsack model for deterministic capital budgeting problem gives the portfolio of projects to choose that maximizes total NPV while staying within the annual budgets. This presented optimization model can be solved by an appropriate optimization software such as CPLEX.[70]

How should we practice capital budgeting when we have uncertain forecasts for these parameters? This is a challenging question, and in order to illustrate how uncertainty can affect a problem in a practical setting, we discuss the following common scenario. Consider that a multiknapsack model has been applied in the capital budgeting process to form a project portfolio. Then, over the course of the year, the available budget reduces owing to some reasons, such as an external event or because one or more projects experience cost overruns. As a specific matter, some other projects are then forced out of the portfolio; that is, they are not carried out.

As solutions, one methodology is to resolve a multiknapsack model when refined predictions for costs, profits, and budgets become available, and another approach is to formulate a developmental model. In the next section, we will deal with an extended model that is proposed by Koc et al. and is applied by various project managers and manufacturing planners.[71]

Set of Notation and Multiknapsack Formulation for the Stochastic Capital Budgeting Problem: The deterministic capital budgeting model assumes that we know the problem data (i.e., project costs and revenues and the budget) in advance with certainty. In this section, an extended model will be presented that explicitly incorporates multiple budget, cost, and revenue scenarios. The need to deal with these uncertain parameters motivates extending the previous model to form a priority list with the goal of maximizing the expected NPV of the projects we can implement after the uncertain parameters are revealed. The notation and formulation of the optimal prioritization model are as follows:

Indices and Sets:

$i, i' \in I$: Candidate projects

$p \in P$: Priorities, P = {1,2, …, |*I*|}

$t \in T$: Candidate periods (years)

$\varpi \in W$: Senarios

Data:

a_i^{ω} Net present value of project *i* under scenario ϖ

b_t^{ω} Available budget in year *t* under scenario ϖ

$c_{i,t}^{\omega}$ Cost of project *i* in year *t* under scenario ϖ

q^{ω} Probability of scenario ϖ

Decision variables:

x_i^{ω} 1 if project *i* is selected under scenario ω; 0 otherwise

$y_{i,It}$ 1 if project *i* has higher priority than project *i'*; 0 otherwise

$z_{i,p}$ 1 if project *i* is assigned priority level *p*; 0 otherwise

Formulation:

Objective function:

$$\max_{x,y,z} \sum_{\omega \in W} q^{\omega} \sum_{i \in I} a_i^{\omega} x_i^{\omega} \tag{38}$$

Subject to :

$$\sum_{i \in I} c_{i,t}^{\omega} \mathrm{x}_i^{\omega} \leq b_t^{\omega} \quad t \in T \ , \ \ \omega \in W \tag{39}$$

$$\sum_{i \in I} z_{i,p} = 1 \qquad p \in P \tag{40}$$

$$\sum_{p \in P} z_{i,p} = 1 \qquad i \in I \tag{41}$$

$$|P| y_{i,i'} \geq \sum_{p \in P} \left(|P| - p \right)(z_{i,p} - z_{i',p}) \qquad i \neq i',\ i,i' \in I \tag{42}$$

$$y_{i,i'} + y_{i',i} = 1 \qquad i < i',\ i,i' \in I \tag{43}$$

$$x_i^{\omega} \geq x_{i'}^{\omega} + y_{i,i'} - 1 \qquad \omega \in W, i \neq i',\ i,i' \in I \tag{44}$$

$$x_i^{\omega} \in \{0,1\} \qquad i \in I,\ \omega \in W \tag{45}$$

$$y_{i,i'} \in \{0,1\} \qquad i \neq i',\ i,i' \in I \tag{46}$$

$$z_{i,p} \in \{0,1\} \qquad i \in I,\ p \in P \tag{47}$$

This model is a two-stage stochastic integer program. The first-stage decision variables, z and y, form the priority list and establish the precedence between projects based on their respective ranking in the list. The second-stage decision variable, x, chooses the portfolio of projects to implement under each scenario. The objective function captures the expected NPV, forming the weighted sum of NPVs over all scenarios.

The first constraint certifies that the implemented projects stay within budget under each scenario, for each year. The second constraint (Eq.36) assigns exactly one project to each priority level, and the third constraint (Eq.37) assigns exactly one priority level to each project. These constraints assume that all projects are assigned a priority (i.e., |P| = |I|). Given z, the fourth and fifth constraints define y, making sure that each pair of projects is properly ordered. The sixth constraint requires that the projects implemented by xϖ, under each scenario, are consistent with the priority list's ordering. The last three sets of constraints are binary restrictions; for instance, either a project is chosen for implementation or not under scenario ϖ (variable xi^{ϖ}). The timing of decisions and observations of uncertainty are key elements in understanding the optimal prioritization of this proposed expanded model. First, the priority list is formed by y and z. Next, the values of the budget, cost, and profit parameters (a^{ϖ}, b^{ϖ}, c^{ϖ}) are identified. We then efficiently work down the priority list of executing the projects, applying x^{ϖ}, until the budget is completely used.

3.4. Process Optimization

Process optimization is the field of regulating a process so as to optimize some definite set of parameters without violating several constraints. The most usual objectives are minimizing cost, maximizing throughput, and enhancing the efficiency of a process. Process optimization is one of the major quantitative techniques in industrial decision making. The actual driving force for process optimization is efficiency. Manufacturing systems comprehend that if they can function more efficiently, it will enhance their bottom line. Environmental issues and reducing natural resources are among the most vital issues in tactical manufacturing decisions. Public awareness and government strategies and regulations have pushed numerous manufacturing systems to focus more on sustainable manufacturing. The major target of sustainable manufacturing is to preserve energy and natural resources more efficiently ϖ. The demand for a better performance and a more economical manufacturing process is integrated with the need for an energy- and resource-efficient production so as to reduce cost and at the same time to abide by the environmental aspects. Once implementing manufacturing process optimization, defining the objective function is a vital issue. Traditionally, maximizing the profit or minimizing the manufacturing cost is the usual objective, subject to meeting quality requirements.[72]

The preferable performance and the energy and resource efficiency can be attained by optimizing the process chain, by improving each process by itself, by enhancing integrated processes, by using more efficient technologies, equipment, and tools as well as by implementing mechatronic control systems alongside the entire production chain. The higher capacity and performance of new machining techniques integrated with a broader technical knowledge of the cutting process have resulted in more economical production and machining of functionally determined surfaces. Such techniques allow more energy-effective use of the machined parts. These days, the more efficient research and development (R&D) process, the prototyping stages, and the optimization of the machining processes count more and more on simulation and modeling techniques, which can save time, costly resources, machining energy, and long-term or pricey investigations.[73] Primary work on machining process optimization with total sustainability as the objective consists of creating a methodology to consider a partial set of indicators generated based on both of the deterministic and nondeterministic elements in sustainable manufacturing processes.[72] In order to implement sustainability efficiently in a manufacturing system, the following suggested aspects of endeavors need to be considered:[74]

- Real-time resource and energy consumption monitoring for better utilization
- Situation-aware and energy-efficient optimization of energy and wastes
- Sustainable manufacturing processes for energy consumption reduction
- Process-oriented dynamics-based energy optimization modeling for process planning and machine/robot control

In the next section, we will deal with various optimization methods and modeling that can be applied to sustainable manufacturing systems in order to achieve more efficient outcomes.

3.4.1. Classification of Optimization Problems Using Mathematical Modeling

From a practical viewpoint, process optimization task refers to finding the best solution for a system or process within the constraints. This task needs the following elements:[75, 76, 77]

- An objective function is required that contributes to a scalar quantitative performance measure needing to be minimized or maximized. This can be the system's cost, yield, profit, and so on.
- A predictive model is needed that identifies the behavior of the system. For the optimization problem this refers to a set of equations and inequalities that are considered constraints. These constraints compose a viable region that determines the limits of performance for the system.

- Variables that appear in the predictive model must be adjusted to satisfy the constraints.

This can usually be implemented with multiple samples of variable values, conducive to a viable area that is defined by a subspace of these variables. In lots of engineering problems, this subspace can be described by a set of decision variables that can be interpreted as degrees of freedom in the process.

In order to search for a proper solution method related to an optimization problem, we need to complete our information in terms of various features, such as the type of constraints, the nature of design variables, the physical structure of the problem, the nature of the equations considered, the deterministic nature of the variables, the permissible value of the design variables, the number of objective functions, and the number of uncertain and probabilistic variables. In the next sections, we present an efficient classification for optimization problems.

3.4.1.1. Classification of Optimization Problems Based on the Existence of Constraints

In classification based on the presence of constraints, there are two main categories: constrained optimization problems, which are bound to one or more constraints, and unconstrained optimization problems, in which no constraints exist.

3.4.1.2. Classification of Optimization Problems Based on the Nature of Design Variables

In classification based on the nature of the design variables, there are two broad classes. The first class accounts for a modeling whose objective is to search for a set of design parameters that make a prescribed function of these parameters minimums or maximums subject to certain constraints. The second category refers to a model in which the objective is to discover a set of design parameters (which are all continuous functions of some other parameter) that minimizes an objective function bounded to a set of constraints.

3.4.1.3. Classification of Optimization Problems Based on the Physical Structure

Based on the physical structure, optimization problems can be categorized as optimal control and nonoptimal control problems. An optimal control problem is a mathematical programming problem consisting of various stages where each stage evolves from the earlier stages in a specified manner. An optimal control problem is described by two types of variables: the control, or design, variables and state variables. Here the objective is to search for a set of control or design variables in a way that the total objective function (also called the performance index) over entire stages is minimized bounded to a set of constraints on the control and state variables. An optimal control model can be formed as follows:

Finding X, which minimizes:

$$f(X) = \sum_{i=1}^{l} f_i(x_i, y_i) \tag{48}$$

Subject to:

$$q_i(x_i, y_i) + y_i = y_{i+1} \qquad i = 1, 2, \ldots, l \tag{49}$$

$$g_j(x_j) \le 0 \qquad j = 1, 2, \ldots, l \tag{50}$$

$$h_k(y_k) \le 0 \qquad k = 1, 2, \ldots, l \tag{51}$$

Where x_i is the ω control variable, y_i is the ith state variable, and f_i is the contribution of the ith phase to the total objective function. g_j, h_k, and q_i are the functions of x_j, y_k , and x_i, y_i accordingly. The parameter of l is the total number of stages. The control and state variables x_i and y_i can be vectors in several cases.

The problems that are not optimal control problems are named nonoptimal control problems.

3.4.1.4. Classification of Optimization Problems Based on the Nature of Expressions for the Objective Function and the Constraints

In accordance with the nature of expressions for the objective function and the constraints, optimization problems can be categorized as linear, nonlinear, geometric, and quadratic programming problems.

Linear Programming: Linear programming refers to a model in which objective function and all the constraints are linear functions of the design variables. A linear programming model can be shown as follows:

$$\text{Finding X} = \begin{Bmatrix} x_1 \\ x_1 \\ . \\ . \\ x_n \end{Bmatrix}, \text{which minimizes } f(X) = \sum_{i=1}^{n} c_i x_i \tag{52}$$

Subject to:

$$\sum_{i=1}^{n} a_{ij} x_i = b_j \qquad j = 1, 2, \ldots, m \tag{53}$$

$$x_i \geq 0 \qquad j = 1, 2, \ldots, m \tag{54}$$

where c_i, a_{ij}, and b_j are constants.

Nonlinear Programming: If any of the functions among the objectives and constraint functions is nonlinear, the problem is named a nonlinear programming problem. Nonlinear optimization is perhaps the largest class of optimization problems. Generally, these problems are very hard to solve. One of the reasons for which nonlinear problems are hard to solve is the existence of local optimums. When trying to minimize general nonlinear functions, applied algorithms may be trapped in local optimums rather than global optimums.

Geometric Programming: A geometric programming (GP) is a type of mathematical optimization problem typified by objective and constraint functions that have a specific form.

Let $x_1, \ldots, x_n$ account for n real positive variables, and $x = (x_1, \ldots, x_n)$ is a vector with components x_i. A real valued function f of x, with the following form (Eq. 55) is named a monomial function or, more easily, a monomial (of the variables $x_1, \ldots, x_n$). The constant c is the coefficient of the monomial, and the constants of $a_1,\ldots,a_n$ are the exponents of the monomial.

$$f(x) = cx_1^{a_1} x_2^{a_2} \ldots x_n^{a_n} \tag{55}$$

where $c > 0$ and $a_i \in R$

Any positive constant is a monomial, as is any variable. Monomials are closed under multiplication and division: If f and g are both monomials, then so are fg and f/g (This contains scaling by any positive constant.) A monomial elevated to any power is also a monomial:

$$f(x)^{\gamma} = \left(cx_1^{a_1} x_2^{a_2} \ldots x_n^{a_n}\right)^{\gamma} = c^{\gamma} x_1^{\gamma a_1} x_2^{\gamma a_2} \ldots x_n^{\gamma a_n} \tag{56}$$

The term *monomial,* as applied here (in the context of GP) is close to, but differs from, the standard concept of *monomial* applied in algebra. In algebra, a monomial has the form as Equation 55, but the exponents a_i must be non-negative integers and the coefficient c is one. Here *monomial* will refer to the definition given above, wherein the coefficient can be any positive number and the exponents can be any real numbers, consisting of negative and fractional. A sum of one or more monomials is shown in a form of a function as Equation 53.

$$f(x) = \sum_{k=1}^{K} c_k x_1^{a_{1k}} x_2^{a_{2k}} \ldots x_n^{a_{nk}} \tag{57}$$

Where $c_k > 0$

The above fucntion is called a posynomial function or, more easily, a posynomial (with K terms, in the variables $x_1, \ldots, x_n$). The term *posynomial* is meant to recommend an integration of "positive" and "polynomial." Any monomial is also a posynomial. Posynomials are closed under addition, multiplication, and positive scaling. Posynomials can be divided by monomials (with the result also a posynomial): If f is a posynomial and g is a monomial, then f/g is a posynomial. If γ is a non-negative integer and f is a posynomial, then $f\gamma$ always makes sense and is a posynomial (as it is the product of γ posynomials). Considering the aforementioned points, standard format for a GP is as follows:

Minimizing $f_0(x)$
Subject to:

$$f_i(x) \leq 1, \ i = 1, 2, \ldots, m \tag{58}$$

$$g_i(x) = 1, \ i = 1, 2, \ldots, p \tag{59}$$

Where f_i are posynomial functions, g_i are monomials, and x_i are the optimization variables. (There is an imbedded constraint that the variables are positive, i.e., $x_i > 0$.) In a standard format of GP, the objective should be posynomial and it should be minimized; the equality constraints can only have the form of a monomial equal to one. In addition, the inequality constraints can only have the form of a posynomial less than or equal to one.[78]

Quadratic Programming: A quadratic programming problem works very well with a nonlinear programming problem consisting of a concave quadratic objective function (maximization problem) and linear constraints. It is typically formulated as follows:

Assume x belongs to R^n space. The $n \times n$ matrix Q is symmetric, and c is any $n \times 1$ vector.

$$\text{Minimize } \tfrac{1}{2} X^T QX + c^T X \text{ (with respect to } X\text{)} \tag{60}$$

where X^T denotes the vector transpose of X.
Subject to one or more constraints of the form: $AX \leq b$

The notation $AX \leq b$ shows that every entry of the vector AX is less than or equal to the equivalent entry of the vector b.

If the matrix Q is a positive semi-definite matrix, then f is a convex function. In such a situation the quadratic program has a global minimizer if there are some feasible X vectors that satisfy the defined constraints and if f is restricted below on the viable region. If the matrix Q is positive definite and the problem has a feasible solution, then the global minimizer is unique. If Q is zero, then the problem becomes a linear program. If there are only equality constraints, the quadratic programming can be solved by a linear system. If not, various methods, such as interior point, active set, conjugate gradient, gradient projection, and extensions of the simplex algorithm, will be applied to solve a quadratic programming model. Convex quadratic programming is also a special case of the more general field of convex optimization.

3.4.1.5. Classification of Optimization Problems Based on the Permissible Values of the Decision Variables: Integer and Real-Valued Programming

Under this classification, the problem can be categorized as integer and real-valued programming problems. If some or all of the design variables of an optimization problem are limited to having only integer (or discrete) values, the problem is named an integer programming problem.

A real-valued problem is that in which it is sought to minimize or maximize a real function by systematically selecting the values of real variables from within an allowed set. When the allowed set contains only real values, it is called a real-valued programming problem.

3.4.1.6. Classification of Optimization Problems Based on the Deterministic Nature of the Variables: Deterministic and Stochastic Programming

Under this classification, optimization problems can be classified as deterministic and stochastic programming problems. In deterministic programming, all the design variables are deterministic, while in stochastic programming, some or all the parameters (design variables and/or pre-assigned parameters) are probabilistic (nondeterministic or stochastic). Stochastic programming models benefit from the fact that probability distributions controlling the data are recognized or can be approximated; the objective is to search for some policies that are feasible for all, or almost all, the possible data examples and optimize the anticipated performance of the model. For example, demands for a different range of products produced by a manufacturing system can have a specific distribution with a known mean and unknown/known state deviation. The rate of producing defective products and produced waste can be considered as stochastic variables. Reliability of machinery can also be considered as a probabilistic variable.

3.4.1.7. Classification of Optimization Problems Based on the Number of Objective Functions: Single- and Multi-Objective Programming

Under this classification, objective functions can be categorized as single- and multi-objective programming problems. There is only a single objective in single-objective programming, and there are multiple objectives in multi-objective programming. Multi-objective optimization can be applied in the optimization process for sustainable manufacturing systems, where optimal decisions must be made in the presence of trade-offs between two or more conflicting objectives. Minimizing costs and waste while maximizing service levels, and maximizing performance while minimizing fuel consumption and greenhouse gas emissions in a manufacturing system, are samples of multi-objective optimization problems consisting of two and three objectives, correspondingly. For a nontrivial multi-objective optimization issue, there is not a single solution that concurrently optimizes each objective. In that case, the objective functions are supposed to be conflicting, and there exists a (possibly infinite) number of Pareto optimal solutions. Pareto optimal, also known as nondominated or noninferior, refers to a situation in which none of the objective functions can be improved in value without degrading some of the other objective values. Lacking additional subjective preference information, all achieved Pareto optimal solutions are possibly better than having no solutions (as vectors cannot be ordered completely), but it is required to search for a set of Pareto optimal solutions, and/or measure the trade-offs in meeting various objectives, and/or search for a single solution that meets the subjective decision maker's preferences, or apply different recommended effective solution methods to obtain optimal values for a developed optimization model.[79, 80]

3.4.1.8. Classification of Optimization Problems Based on Modeling Under Uncertainty

Uncertainty usually exists in real-world applications. Information applied to model a problem can be noisy and imperfect. Conventionally, sensitivity analysis was conducted to study the effect of perturbations on particular designs. Whereas these methodologies can be applied to compare various designs, they do not inherently search for one that is less sensitive, namely, they do not enhance the robustness directly. Stochastic optimization is the traditional method to address optimization under uncertainty. This method takes a probabilistic approach. The probability distribution of the uncertainties is approximated and combined with the model by applying chance constraints, risk measures, and numerous scenarios emulating the distribution, but in some cases the actual distribution of the uncertainties is not accessible. Any particular probability distribution is, at best, a specialist's opinion. Other methodologies incorporate uncertainties and perturbations through tolerance bands and margins in the relevant multi-objective function as taking constraints into consideration by adding a penalty term to the initial constraints.[81]

Robust optimization, which is also a relatively new method to dealing with optimization issues influenced by uncertainty, has by this time proven so helpful in real applications that it is hard to cope with such problems today without using this powerful approach. Robust optimization is developed to satisfy several main issues related to uncertainty-affected optimization problems, such as operating under lack of full information on the nature of uncertainty, modeling the problem in a form that can be solved effectively, and providing assurances about the performance of the solution.

Different decision-making models under uncertainty suppose complete information, for instance, precise values for the system parameters and specific probability distributions of the random variables. In reality, such an accurate knowledge is seldom available in practice, and thus a strategy according to inaccurate inputs may be infeasible or display poor performance when implemented. Considering such a situation, stochastic optimization begins with the assumption that the uncertainty has a probabilistic description. In robust optimization, random variables are considered as uncertain parameters belonging to a convex uncertainty set and the decision maker supports the system against the worst case in that set. Robust optimization is specifically different from sensitivity analysis, which is usually used as a post-optimization technique to measure the modifications in the cost of the related optimal solution with minor perturbations in the underlying problem data. In robust optimization, the objective is to calculate fixed solutions that make sure about viability independent of the data. In other words, these solutions have a deductive ensured viability when the problem parameters change in the prescribed uncertainty set, which may be sizable.[82, 83, 84]

3.4.2. Using Statistical Methods for Process Optimization

Statistical methods are categorized among effective tools for process optimization especially when it is essential to recognize main factors and interactions in a process or use sampling to control a system. In the next section, we concentrate on two essential techniques for industrial production processes where the main emphasis is process optimization: statistical design of experiments and principal component analysis.

3.4.2.1. Statistical Design of Experiments

Statistical design of experiments (DoE) is a powerful tool for optimizing processes. In manufacturing systems, DoE is recognized as an effective statistical method that can be applied for numerous experimental examinations. Every machine applied in a production process allows its operators to modify different settings, having an effect on the subsequent quality of the product produced by the machine. Regarding this situation, experimentation helps the production engineer to adjust the settings of the machine in an organized manner and to discover which factors have the greatest impact on the subsequent quality. Applying this information, the settings can be continuously improved until optimum quality is achieved. DoE technique is one of the powerful tools utilized in order to evaluate the causes of process variation and reduce the number of manufactured defective parts or products. Major concepts in developing a designed experiment contain blocking, randomization, and replication. A well-performed experiment may offer answers to questions such as:

- What are the key factors and their related levels in a process?
- Which structure and layout of experimental runs should be considered?
- In what settings would the process deliver satisfactory performance?

- What are the major interaction effects in the process?
- What settings would result in less variation in the output?

The objective of experimental methods is to recognize the optimum settings for the various factors that have an effect on the production processes. The design of experiments is an organized method to decide the relation between factors having effects on a process and the output of that process. In other words, it is applied to search for cause-and-effect relationships.

This information is required to manage process inputs to improve the output. For understanding the DoE, we need to have knowledge of some statistical tools and experimentation concepts. A DoE can be evaluated in many software programs, such as DOE++: a software tool for experiment design and analysis, Minitab, SPSS, Design-Ease® software, and Design-Expert® software; however, it is vital for a specialist to comprehend basic DoE concepts for appropriate application.

The most typically applied terms in the DoE method can be categorized as controllable and uncontrollable input factors, responses, hypothesis testing, blocking and replication, and interaction.

- Controllable input factors, or x factors, are input parameters that can be changed in an experiment or process.
- Uncontrollable input factors are parameters that cannot be changed. These factors require to be identified to know how they may have an effect on the response. In addition, uncontrollable factors that engender variation under normal operating situations are signified as noise factors. Noise factors such as multiple machines, multiple shifts, raw materials, humidity, and so forth can be built into the experiment so that their variation does not get lumped into the unexplained, or experimental error. A major power of designing experiments is the ability to determine major factors and the related settings that minimize the impacts of the uncontrollable factors.
- Responses or output measures are the components of the process outcome that gauge the desired effect. The controllable input factors can be revised to optimize the output. The relation among the factors and responses is demonstrated by Figure 3.31.
- Hypothesis testing assists to identify the key factors, by applying statistical methods.

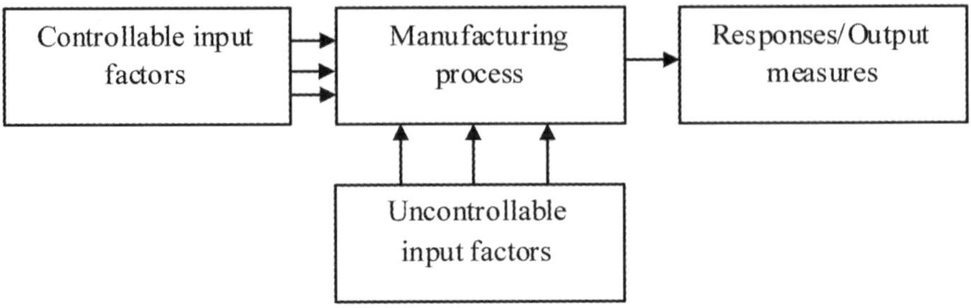

Figure 3.31. Process Factors and Responses

- Blocking is an experimental method to prevent any useless variations in the input or experimental process. As an example, an experiment may be performed with the same equipment to avoid any equipment variations. Specialists also replicate experiments, conducting the same combination run more than once, to achieve an approximation for the amount of random error that could be part of the process.
- An interaction is a condition in which the concurrent effect of two variables on a third is not an additive term, once an experiment has three or more variables.

Some major categories of designs that are usually applied in industrial experimentation can be classified as follows:

- $2^{(K-P)}$ (two-level, multi-factor) fractional factorial designs (focusing on the concept of design resolution, Plackett-Burman (Hadamard Matrix) designs for screening, improving design resolution via fold-over, aliases of interactions: design generators, blocking, replicating the results, partial replication, adding center points.
- $2^{(K-P)}$ (two-level, multi-factor) maximally un-confounded and minimum aberration designs.
- $3^{(K-P)}$ (three-level, multi-factor) designs.
- Central composite and nonfactorial response surface designs. The response surface methodology is also a statistical modeling tool, which applies the

regression analysis to develop the relationship between different process parameters and response.

- Latin square designs.
- Taguchi robust designs.
- Mixture designs and triangular surfaces.
- Designs for constrained surfaces and mixtures.
- Creating D- and A-optimal designs for surfaces and mixtures.

3.4.2.2. A Multivariate Statistical Method: Principal Component Analysis

Principal component analysis (PCA) is a data reduction technique applied to recognize a small set of variables that account for a large portion of the total variance in the initial variables. The major objectives of a PCA are the analysis of data to identify patterns and discovering trends to reduce the dimensions of the data set with minimal loss of information. A PCA is essential to the study of multivariate data. This method is also utilized to recognize "latent" dimensions in the data. In effect, a PCA computes linear combinations of variables. The first linear combination of variables accounts for the largest amount of variation in the sample. The second combination explains the next largest amount of variance in a dimension independent of the first, and so on. To monitor the progress of batch processes in a manufacturing system, a multivariate statistical procedure can be applied. Such a procedure assists to reduce the amount of defective products and waste by utilizing a historical database of earlier successful batches.[85] Listed below briefly summarizes the six typical stages for implementing a PCA:

1. Taking the whole data set consisting of d-dimensional samples disregarding the class labels
2. Computing the d-dimensional mean vector (for example, the means for every dimension of the whole data set)
3. Calculating the scatter matrix (alternatively, the covariance matrix) of the whole data set
4. Computing eigenvectors and corresponding eigenvalues

5. Sorting the eigenvectors by reducing eigenvalues and choosing k eigenvectors with the largest eigenvalues to form a $d \times k$ dimensional matrix W (every column represents an eigenvector)
6. Applying the $d \times k$ eigenvector matrix to transform the samples onto the new subspace

This can be reviewed by the following equation:

$$y = W^T \times x \tag{61}$$

(where x is a dimensional vector referring to one sample and y is the transformed dimensional sample in the new subspace).

3.4.3. Lean Manufacturing: Methodologies and Their Application in Process Optimization

Lean methodology as a process management philosophy has speedily obtained recognition in the manufacturing sectors throughout recent years. Lean manufacturing is a systemic method for the elimination of waste within a manufacturing process. Lean manufacturing concentrates on waste removal from within the facility's production systems through continuous improvement and process modifications for reducing nonvalue-added activities or elimination of wastes. Lean manufacturing methodologies also take into account waste created through overburden and waste created through unevenness in workloads. The basis of "lean" can be traced back to Toyota in the 1950s, and it was completely recognized by the 1980s with the publication of *The Machine That Changed the World*. In the1990s, the lean concept was not only extended to various industries but also to much more than simply a "Toyota" and "Tools" orientation to a more inclusive philosophy containing both strategic and organizational components.

Lots of empirical studies have indicated that an implementation of lean methodologies can lead to better outcomes. However, in some extreme cases, manufacturing systems may become almost preoccupied by cost and/or staff reduction that the work becomes more intense and stressful. In other words, some manufacturing systems have problems in sustaining lean over time. One description of the troubles manufacturing systems face in sustaining lean may be accredited to a lack of concentration on the developmental progression of lean capabilities among the members of the system. By concentrating on developing lean capabilities, members of the system should then become increasingly better at doing lean while at the same time making a learning environment that advocates a lean culture.[86]

Lean manufacturing is a multidimensional method that entails a broad variety of management practices, such as just-in-time (JIT) production, total quality control (TQC), cellular manufacturing, and supplier management in a combined system.[87] On the whole, review of associated research verifies that implementation of lean practices is often related to enhancements in operational performance measures. The most usually cited advantages associated with lean practices refer to the enhancement in labor productivity and quality, along with the reduction in customer lead time, cycle time, wastes, and manufacturing costs.

The following is a group of 25 fundamental lean tools that can be applied in sustainable manufacturing systems:

- **5S (Sort, Set in order, Shine, Standardize, and Sustain):** Here *Sort* means removing items that are not required; *Set in order* refers to arranging remaining items; *Shine* refers to cleaning and investigating the work area; *Standardize* means to develop standard daily practices; and *Sustain* denotes regularly applying the standards. This lean tool organizes the work area in a manufacturing system and removes waste that stems from a poorly organized work area (e.g., wasting time in search of a tool).
- **Andon:** It signifies a visual feedback system for the facility floor that specifies production status, alerts when assistance is required, and empowers operators to halt the production process. It acts as a real-time communication technique for the plant floor that brings prompt attention to issues as they happen; hence, they can be promptly addressed.
- **Bottleneck analysis:** This kind of analysis recognizes which parts of the manufacturing process to limit the overall throughput. It also improves the performance of that part of the process. This tool enhances throughput by fortifying the weakest link in the manufacturing process.
- **Continuous flow:** The objective of this methodology refers to manufacturing where work-in-process smoothly flows through production with minimal or no buffers between various stages of manufacturing processes. It removes various forms of waste, such as inventory wastes and waiting time spans.

- **Gemba (the real place):** Gemba refers to a philosophy that rings a bell for managers and production engineers to get out of offices and spend time on the plant floor, the place where the real action happens. It contributes to a deep and thorough understanding of real-world manufacturing issues by firsthand observation and by talking with plant floor employees.
- **Heijunka (level scheduling):** It refers to a form of production scheduling that knowingly manufactures in much smaller batches by sequencing product variants within the same process. It reduces lead times, as each product or variant is manufactured more often. It also decreases inventory, since batches are smaller.
- **Hoshin kanri (policy deployment):** This method tries to align the objectives and strategies of a manufacturing system, with the plans and tactics of middle management tactics and the work performed on the plant floor. It makes sure that movement toward strategic objectives is consistent and thorough.
- **Jidoka (automation):** In this technique, equipment is designed in a way to moderately automate the manufacturing process and to automatically halt when defects are detected. After jidoka, workers can often monitor multiple stations (reducing labor costs) and various quality issues can be detected instantly (improving quality).
- **Takt time:** Takt time is the average unit production time required to satisfy a customer's demand. This is also a highly efficient method in reducing inventory levels. It enhances cash flows and reduces space requirements. It is computed as planned production time/customer demand and offers a simple, consistent, and intuitive technique of pacing production. It can be developed to set an efficiency goal for the plant floor (actual pieces/target pieces).
- **Just in time (JIT):** This method pulls components and parts through production in accordance with customer's demand, rather than pushing parts through production based on the forecasted demand. JIT counts on various lean tools, such as continuous flow, heijunka, kanban, standardized work, and takt time.
- **Kaizen (continuous improvement):** This is a tactic where employees work together proactively to achieve consistent, incremental improvements in the manufacturing process. It integrates the cooperative capabilities of a manufacturing system to develop an engine for persistently removing waste from manufacturing processes.

- **Kanban (pull system):** Kanban is a technique of adjusting the flow of products and materials, both inside the factory and with outside suppliers and customers, by using automatic replenishment through signal cards that designate when more substances, materials, or products are required. It removes waste from inventory and overproduction. It can also remove the need for holding more physical inventories. In this technique, signal cards are applied in order to specify in which periods more materials are required and when ordering should occur.
- **KPI (key performance indicator):** This represents metrics that are designed to be applied for tracing performance in a manufacturing system based on the defined major objectives. These metrics are aligned with top-level strategic objectives, thus helping to achieve those objectives. They are efficient in exposing and measuring waste. Moreover, they are instantly affected by plant floor employees; thus they can drive outcomes.
- **Muda (waste):** This refers to whatever does not add value, from the customer's viewpoint, in the manufacturing process. Removing waste is the core concentration of lean manufacturing.
- **Overall equipment effectiveness (OEE):** It is a method for measuring productivity loss for a manufacturing process. Three classes of loss that are traced consist of availability such as down-time, performance such as slow cycles, and quality such as percentage of rejects. It presents a benchmark/baseline and a means to trace improvements in removing waste from manufacturing processes. With this method, 100% OEE denotes a perfect production system that manufactures only good parts, as fast as possible, and with no downtime.
- **PDCA:** This is an iterative methodology for fulfilling improvements with various steps:

    1. Plan (establish a plan, establish a hypothesis, and gather expected results).
    2. Do (implement a plan and run experiments).
    3. Check (verify and analyze expected results achieved).
    4. Act (review and assess; revise the experiment and do it again).

- **Poka-yoke (error proofing):** This is a method for designing error detection with the goal of achieving zero defects in production systems. It is hard and costly to discover all defects through inspection, and amending defects typically get substantially more expensive at each phase of production.

- **Root cause analysis:** RCA is a problem-solving method that concentrates on resolving the underlying issue rather than using rapid fixes that deal with only the immediate symptoms of the issue. A common procedure is to ask why five times, each time moving a step closer to discovering the true underlying problem. It assists to make sure that a problem is truly removed by utilizing corrective actions to the "root cause" of the problem.
- **Single-minute exchange of dies (SMED):** This technique reduces setup/changeover time periods and entails the following process:
    1. Converting setup steps to be external (performed while the process is running).
    2. Simplifying the internal setup (e.g.,, changing bolts with knobs and levers).
    3. Removing nonessential operations.
    4. Creating standardized work instructions: Using standardized work instructions in a manufacturing system is accompanied by practical advantages such as manufacturing in smaller lots, reducing inventory, and enhancing customer responsiveness.
- **Six big losses:** Six classes of productivity loss that are experienced commonly in manufacturing systems are as follows:
    1. Breakdowns.
    2. Setup/Adjustments.
    3. Small stops
    4. Reduced speed.
    5. Startup rejects.
    6. Production rejects.

The six big losses method offers a framework for targeting the most common causes of waste in manufacturing systems.

- **SMART goals:** These are *s*pecific, *m*easurable, *a*chievable, *r*ealistic, and *t*imely goals
- **Standardized work:** This represents documented practices for manufacturing systems that explain best procedures containing the time to complete

each task. Standardizing various tasks of processes helps to remove waste consistently and form a baseline for upcoming improvement tasks.

- **Total productive maintenance (TPM):** This is a comprehensive methodology related to maintenance that emphasizes proactive and preventive maintenance to optimize the operational time of equipment. This methodology focuses on greater involvement of plant floor workers and managers in terms of accomplishing TPM programs' goals and objectives. This technique can result in productivity improvement, reducing cycle times, and eliminating defects.

- **Value stream mapping:** This tool is applied to visually map the flow of production. It demonstrates the current and future status of processes in a way that underlines opportunities for improvement. It exposes waste in the current processes and offers a road map for improvement. Lean value stream manufacturing is also based on this tool, and it refers to the incorporation of lean value concepts and design for sustainability.

- **Visual factory:** Visual indicators, displays, and controls are applied all over manufacturing sites to enhance communication of information. This makes the status of manufacturing processes simply accessible and very evident to managers and practitioners.

Applying lean manufacturing methodologies in sustainable manufacturing systems can protect the environment by reduction of waste of raw materials and energy supplies in a very effective way. For instance, lean value stream manufacturing can be taken into account as a life cycle–based endeavor in order to enhance the life cycle of the products in a manufacturing system. From a managerial standpoint, lean manufacturing and environmental management practices are synergistic in terms of their emphasis on eliminating waste and inefficiency. The focus of lean manufacturing on internal and process waste reduction can be expanded with an emphasis on environmental waste management strategies. In some cases, environmental management practices do demand additional resource investments. It is better for manufacturing firms to implement both lean manufacturing and environmental management practices even at the same time to take advantage of accompanied improvements regarding their system's performance. This will also allow firms to satisfy their sustainable manufacturing performance objectives effectively.[88]

A sample road map for a sustainable remanufacturing system that takes integrated methodologies such as lean manufacturing, failure modes and effects analysis (FMEA), and six sigma tools into consideration is shown in Figure 3.32. The term *six sigma processes* originates from the notion that if a process has six standard deviations between the process mean and the closest specification limit, specifically no product will fail to satisfy specifications. *Six sigma quality* is also a term generally applied to specify a process is well controlled within process limits $\pm 3\ \sigma$ (sigma is a true process standard deviation value) from the center line on a control chart, and requirements/tolerance limits $\pm 6\ \sigma$ from the center line.[89]

As production and manufacturing are resource-demanding and costly endeavors, identifying a proper methodology to incorporate lean, green, and sustainable tactics into a manufacturing system can be conducive to production cost reductions.

In this section, we describe an evaluation method for simulating, optimizing, and valuating a manufacturing system's performance indicators while applying the impacts of implementing an integration of lean, green, and sustainability practices and strategies in a manufacturing facility. To set up a simulation model of a manufacturing system and successively optimize it, various data demand to be gathered. The required data for each manufactured product can be categorized into operational and production data, energy and resource requirements, and employee involvement. All data needed to establish the discrete event simulation model of the current status can be measured on the shop floor or approximated with historical data. With the purpose of creating a simulation model capable of valuating the increase in a production system's performance, installing multiple levels and constructing reusable modules are recommended. In a regular discrete production environment, one level shows the value stream while another level reveals machine operations (e.g., tool changes, inspections, or setups). The value stream and machine operations can be interrelated by combining inputs from equipment operators or other staff in the production area.

After collecting the data dealing with the manufacturing systems' present status, lean, sustainable, and green tactics must be chosen for the assimilation into the simulation model and its successive analyses, which can also be performed by benchmarking or consulting with specialists.

The aforementioned stages, defined in accordance with maturity models for processes, are utilized to recognize the manufacturing system's improvement in using lean, sustainable, and green tactics: (1) initial, (2) planned, (3) defined, (4) measured, and (5) optimized. The use of stages requires that employees outline the current and desired status of the lean

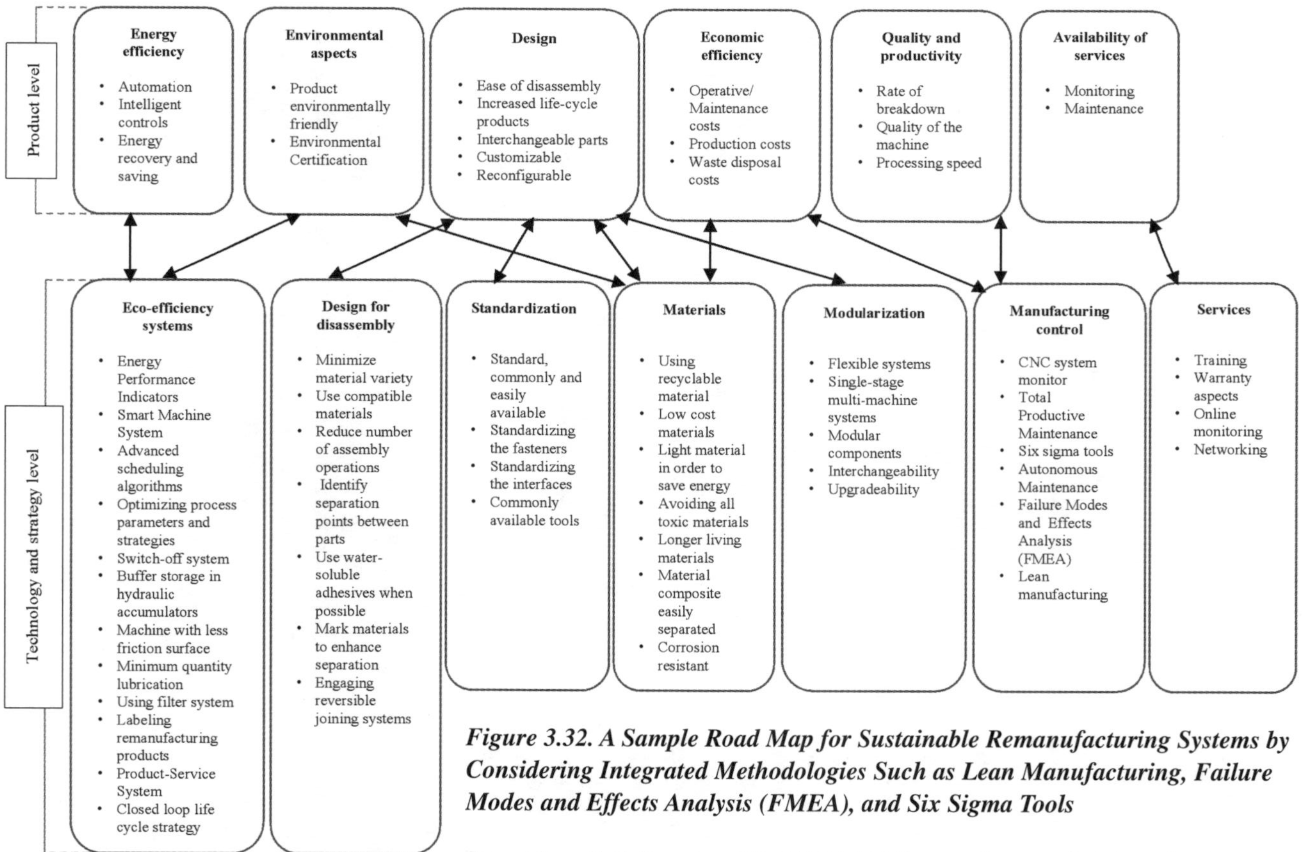

Figure 3.32. A Sample Road Map for Sustainable Remanufacturing Systems by Considering Integrated Methodologies Such as Lean Manufacturing, Failure Modes and Effects Analysis (FMEA), and Six Sigma Tools

Sources: Yang, M. G., Hong, P., and Modi, S. B. (2011), "Impact of lean manufacturing and environmental management on business performance: An empirical study of manufacturing firms"; Cunha, V. P., Balkaya, I., Palacios, J., Rozenfeld, H., and Seliger, G., "Development of Technology Roadmap for Remanufacturing Oriented Production Equipment," in Advances in Sustainable Manufacturing.

and green strategy. To designate optimal intervals for the simulation, progression potentials (e.g., the possible reduction in batch size or setup time) for each stage should be given. It may also be accomplished with the help of consultation with a specialist or benchmarking.

In a discrete event simulation model, manufacturing policies may be parameterized with parameter or structural adjustments. Parameter changes are accomplished by changing values in a predefined optimization interval. For instance, consider that in a manufacturing system, setup time was reduced by 9.5% by applying stage 2 (planned) of the lean method: SMED. Related to this manufacturing system, the specialist anticipates that further optimization would result in a 38% reduction in setup time in comparison with its original value without SMED, a 9.5% (38/4 = 9.5) reduction at each stage relative to the former stage. This agrees with an optimal interval of [0.69; 1.00] relative to stage 2. Regarding this interval, particular, the value of 1.00 shows the current value (stage 2), which accounts for the 9.5% improvement, relative to the initial value (stage 1). It is assumed that the range of the interval of 0.31 represents the maximum amount to be saved when optimized (stage 5) relative to the current value (stage 2); therefore, the first term of the given interval is 0.69.

Modeling other strategies may need structural modifications of the simulation model's internal logic. For example, green and sustainable scheduling strives to lower resource consumption by utilizing machine tool scheduling methods. Changes to the machine tool schedule need modifications of the simulation model's parameters impacting the scheduling sequence; hence, a structural modification is needed.

When appropriate lean, sustainable, and green tactics are integrated, the simulation model of the current state can be connected to parameter optimization software to improve the manufacturing system's performance. Here, some of the leanest, sustainable, and green strategies regarding the operational stages of the factory and machines with their expected effect are presented. The required modifications in the parameters used for simulation are also mentioned.[90]

- Strategy: Batch reduction; Impact: Reduction in batch sizes; improvement in lead times (a lead time is the latency (the delay) between the initiation and execution of a process); shorter waiting times; Change to simulation: Parameter for batch sizes will be modified (e.g., [0.1; 1.0] if it can be reduced to 10%).

- Strategy: Change in production control system (e.g., applying kanban system); Impact: Less inventory, improvement in lead times, and shorter waiting times; Change to simulation: Structural changes based on the production control system used are needed.
- Strategy: Defect reduction (e.g., total quality management [TQM]), Impact: Less re-work, less scrap, and power reduction (planned amount of good parts to be produced faster); Change to simulation: Parameter for quality rate will be modified (e.g., [1.0; 1.1] if it can be improved by 10%).
- Strategy: Process integration; Impact: Improvement in transportation, shorter waiting times; Change to simulation: Structural changes within the factory layout (e.g., combined processes) are needed.
- Strategy: Green and sustainable scheduling; Impact: Power reduction, potential effects on waiting and idle time depending on scheduling option chosen, affects make-span, Change to simulation: Structural adjustments influencing the scheduling sequence, order, and machine selection are needed.
- Strategy: Reduction of setup time for machines used (e.g., SMED); Impact: Reduction in internal and external setup times; Change to simulation: Parameter for setup times will be modified (e.g., [0.1; 1.0] if it can be reduced to 10%).
- Strategy: Production lead-time reduction for machines; Impact: Reduction in energy consumption; Change to simulation: Parameter for cycle times will be modified (e.g., [0.1; 1.0] if it can be reduced to 10%).
- Strategy: Process integration regarding machines; Impact: Less tool changeovers, setup time reduction; Change to simulation: Parameter for setup times and amount of tool changes will be modified.
- Strategy: Use of more efficient components regarding operational phase of the machines; Impact: Reduction in energy consumption; Change to simulation: parameters that are related to energy consumptions with change.

Considering the aforementioned strategies, here a cost-minimization sample model is presented. This assessment methodology can also be applied when searching for alternative objectives or a multi-criteria decision system, such as the minimization of energy and the reduction of greenhouse gas emissions. In the following model, the objective function of the cost-minimization problem contains the entire variable costs of the assessed production system (C_p). These costs consist of five summations for all different kinds of product (i): opportunity costs ($C_{i,\ opportunity}$), material costs ($C_{i,\ material}$), machine costs ($C_{i,\ machine}$), inventory costs ($C_{i,\ inventory}$), and personnel costs ($C_{i,\ personnel}$).

$$Min\ C_p = Min \sum_{i=1}^{I} \left\{ C_{i,\ opportunity} + C_{i,\ material} + C_{i,\ machine} + C_{i,inventory} + C_{i,\ personnel} \right\} \quad (62)$$

Subject to:

$$x_1 \in \left[a_1, b_1\right] \quad (63)$$

$$x_2 \in \left[a_2, b_3\right] \quad (64)$$

$$\ldots$$

$$x_9 \in \left[a_9, b_9\right] \quad (65)$$

The above objective function is bounded by the constraints consisting of the lean, sustainable, and green strategies. Each lean, sustainable, or green strategy is considered as a function $f(x)$, with x being either a parameter change (e.g., $x_x \in [a_x, b_x], a_x, b_x \in R$) or a structural change (e.g., $x \in [0, 1]$). Resource and energy consumption are incorporated in material costs and machine costs, correspondingly. Machine costs are the result of multiplying the machine's hourly rate of machine j while pursuing an activity $h(mhr_{j,h})$ with the time of various kinds of product i occupying machine j within the simulation period ($T_{i,j,h}$).[90]

$$C_{i,\ machine} = \sum_{j=1}^{J} \left(\sum_{h=1}^{H} mhr_{j,h} \times T_{i,j,h} \right) \quad (66)$$

Energy costs are accounted for as part of the machine's hourly rates. They can be obtained by summing the costs of electricity, gas, and fossil fuels and multiplying consumption rates with prices and production lead times. The proposed model may be solved by applying heuristics, for example, an evolutionary algorithm and simulation in order to evaluate various scenarios. Moreover, regarding the monetary evaluation of lean, sustainable, and green strategy implementation, applying cost-time profiles can be helpful. By using cost-time profiles, the most advantageous strategies can be recognized that are conducive to efficient action plans for the considered manufacturing system. The origins of cost-time profiling lie in the late Westinghouse Electric Corporation. This process is developed for monitoring time-based investments precisely. A cost-time profile offers a means for understanding the dynamics of investment.

A cost-time profile illustrates the accrued costs that have been expanded throughout the manufacturing processes of a product at every time unit of a process. The area below the graph of a cost-time profile is known as "cost-time investment," as it indicates how much money has been tied up in the manufacturing processes and for how long before being recovered through sales. It is the composition of the cost and time dimensions. The use of the cost-time profile and cost-time investment will enhance the accuracy of the direct cost of products, as it demonstrates a hidden cost that has not been accounted for (the investment cost). The cost-time profile can be utilized to evaluate the expected impact of an adjustment in the production processes. There might be saved and new costs that are not absolutely clear and can be discovered through the creation of a cost-time profile. As an example, the switch to operations that are quicker, but costlier or slower, but more economical might have an unanticipated effect in the cost-time profile and cost–time investment that could make removing options attractive again.[91]

References

1. Burrill, C., and Ellsworth, L., *Modern Project Management: Foundations for Quality and Productivity* (Tenafly, NJ: Burrill-Ellsworth Associates, 1980), 209–223.

2. Warda F. A., Pulido-Velázquezb, M. (2008), "Efficiency, equity, and sustainability in a water quantity–quality optimization model in the Rio Grande basin," *Ecological Economics* 66, 23–37.

3. Belchera, K. W., Boehmb, M. M., and Fultona, M. E. (2004), "Agroecosystem sustainability: A system simulation model approach," *Agricultural Systems* 79, 225–241.

4. "Canada's Greenhouse Gas Emissions Trends, 2013," Environment Canada, www.ec.gc.ca/ges-ghg/, retrieved September 4, 2014.

5. Despeisse, M., Ball, P. D., and Evan S., "Modelling and Tactics for Sustainable Manufacturing: an Improvement Methodology" in *Sustainable Manufacturing: Shaping Global Value Creation*, edited by Günther Seliger (New York: Springer, 2012), 9–16.

6. "Energy Consumption in the UK," Gov.UK, Department of Energy & Climate Change (DECC 2013), www.decc.gov.uk/en/content/cms/statistics/publications/ecuk/ecuk.aspx, retrieved September 10, 2014.

7. Ball, P. D., et al. , "Modelling Energy Flows Across Buildings, Facilities and Manufacturing Operations," International Manufacturing Conference, IMC28 (Dublin, August 30–September 1, 2011).

8. "U.S. GHG Inventory 2014, Executive Summary," Environmental Protection Agency, www.epa.gov/climatechange/Downloads/ghgemissions/US-GHG-Inventory-2014-Chapter-Executive-Summary.pdf, retrieved November 4, 2014.

9. Ness, B., Urbel-Piirsalu, E., Anderberg, S., and Olsson, L. (2007), "Categorizing tools for sustainability assessment," *Ecological Economics* 60, 498–508.

10. Pope, J., Annandale, D., and Morrison-Saunders, A. (2004), "Conceptualizing sustainability assessment," *Environmental Impact Assessment Review* 24, 595–616.

11. Devuyst, D. (2000), "Linking impact assessment and sustainable development at the local level: The introduction of sustainability assessment systems," *Sustainable Development* 8, 67–78.

12. "OECD Sustainable Manufacturing Toolkit," OECD Directorate for Science, Technology and Industry (DSTI), www.oecd.org/innovation/green/toolkit/, retrieved Dec.9, 2014.

13. "The Higg Index," www.apparelcoalition.org/higgindex/, retrieved Dec. 9, 2014.

14. "Dow Jones Sustainability Indices," www.sustainability-indices.com, retrieved December 9, 2014.

15. "S&P, TSX Launch Cleantech Index," SustainableBusiness.com, www.sustainable-business.com/index.cfm/go/news.display/id/20009, retrieved January 12, 2015.

16. Elahi, B., Seyed-Hosseini, S. M., and Makui, A. (2011), "A fuzzy compromise programming solution for supplier selection in quantity discounts situation," *International Journal of Industrial Engineering & Production Research* 22, no. 2, 107–114.

17. Lu, T., Rotella, G., Feng, S. C., Badurdeen, F., Dillon, J. O. W., Rouch, K., and Jawahir, I. S., "Metrics-Based Sustainability Assessment of a Drilling Process" in *Sustainable Manufacturing: Shaping Global Value Creation*, edited by Günther Seliger (New York: Springer, 2012), 59–64.

18. Mertins, K., Kohl, H., and Riebartsch, O. 2012, "Sustainable Key-Figure Benchmarking for Small and Medium Sized Enterprises" in *Sustainable Manufacturing: Shaping Global Value Creation*, edited by Günther Seliger, G. (New York: Springer, 2012), 309–313.

19. Joshi, Girdhar, *Management Information Systems* (New Delhi: Oxford University Press, 2013), 20–55.

20. Evans, L., Lohse, N., and Summers, M. (2013), "A fuzzy-decision-tree approach for manufacturing technology selection exploiting experience-based information," *Expert Systems with Applications* 40, 6412–6426.

21. "Featured Sustainable Manufacturing Tools," www.epa.gov/sustainablemanufacturing/tools.htm, retrieved January 15, 2015.

22. "Sustainability-Based Decision Making," www.epa.gov/nrmrl/std/decision_making.html, retrieved January 15, 2015.

23. Mahmoud, M., Liu, Y., Hartmann, H., Stewart, S., Wagener, T., Semmens, D., Stewart, R., Gupta, R., Damian, D., Dominguez, F., Hulse, D., Letcher, R., Rashleigh, B., Smith, C., Street, R., Ticehurst, J., Twery, M., Delden, H. V., Waldick, R., White, D., and Winter, L. (2009), "A formal framework for scenario development in support of environmental decision-making," *Environmental Modeling & Software* 24, 798–808.

24. Liu, Y., Gupta, H., Springer, E., and Wagener, T. (2008), "Linking science with environmental decision making: Experiences from an integrated modeling approach to supporting sustainable water resources management," *Environmental Modeling & Software* 23, no. 7, 846–858.

25. Azevedo, A., ed., *Advances in Sustainable and Competitive Manufacturing Systems: 23rd International Conference on Flexible Automation and Intelligent Manufacturing*, Vol. 1 (New York: Springer, 2013), 259–270.

26. Muroyama, A. P., Mani, M., Lyons, K., and Johansson, B. (2011), "Simulation and Analysis for Sustainability in Manufacturing Processes," National Institute of Standards and Technology, U.S. Department of Commerce, NISTIR 7790.

27. Gaziulusoy, I., Boyle, C., and McDowall, R. (2013), "System innovation for sustainability: A systemic double-flow scenario method for companies," *Journal of Cleaner Production* 45, 104–116.

28. Jawahir, I. S., *Sustainable Manufacturing: The Driving Force for Innovative Products, Processes and Systems for Next Generation Manufacturing* (Lexington: University of Kentucky, College of Engineering and Institute for Sustainable Manufacturing, 2010), 7–10.

29. Eurostat (2001), "Economy-Wide Material Flow Accounts and Derived Indicators: A Methodological Guide," Statistical Office of the European Union, Luxembourg.

30. Eurostat (2009), "Economy Wide Material Flow Accounts: Compilation Guidelines for Reporting to the Eurostat Questionnaire," Statistical Office of the European Union, Luxembourg.

31. Eurostat (2012), "Economy-Wide Material Flow Accounts (EW-MFA): Compilation Guide," Statistical Office of the European Union, Luxembourg.

32. Azevedo, A., ed., *Advances in Sustainable and Competitive Manufacturing Systems: 23rd International Conference on Flexible Automation and Intelligent Manufacturing*, Vol. 1 (New York: Springer, 2013), 1677–1688.

33. "STAN software for substance flow analysis, deutsch version," www.iwa.tuwien.ac.at, retrieved January 18, 2015.

34. "STAN software for substance flow analysis, english version," http://stan2web.net/, retrieved January 18, 2015.

35. Franchetti, M., *Solid Waste Analysis and Minimization: A Systems Approach* (New York: McGraw-Hill, 2009), 21–75.

36. "Life-Cycle Assessment," www.epa.gov/sustainability/analytics/life-cycle.htm, retrieved January 20, 2015.

37. Dassisti, M., Chimienti, M., Shuaib, M., Badurdeen, F., and Jawahir, I. S., "Sustainable Manufacturing: A Framework for Ontology Development" in *Sustainability Manufacturing: Shaping Global Value Creation,*" edited by Günther Seliger (New York: Springer, 2012), 33–39.

38. Solding, P., and Gullander, P. (2009), "Concepts for Simulation Based Value Stream Mapping" in *Proceedings of the 2009 Winter Simulation Conference,* edited by M. D. Rossetti, R. R. Hill, B. Johansson, A. Dunkin, and R. G. Ingalls (Piscataway, NJ: Institute of Electrical and Electronics Engineers, 2009), 2231–2237.

39. EPA (2007), "Lean and Environment Toolkit," U.S. Environmental Protection Agency, www.epa.gov/lean/toolkit/LeanEnviroToolkit.pdf.

40. *Purchase, Operating and Maintenance of Production Equipment Using Life Cycle Costing (LCC),* VDI 2884 (Berlin: VDI, 2005).

41. Dornfeld, D. A. *Green Manufacturing: Fundamentals and Applications* (New York: Springer, 2013), 49–81.

42. Life Cycle Management: "How Business Uses It to Create Opportunities and Make Value Chains More Sustainable," United Nations Environment Programme, 2007.

43. Rebitzer, G., and Hunkeler, D. (2003), "Life cycle costing in LCM: Ambitions, opportunities, and limitations—discussing a framework," *International Journal of Life Cycle Assessment* 8, no. 5, 253–256.

44. Yuan, C. Y., Zhang, T., Rangarajan, A., Dornfeld, D., Ziemba, B., and Whitbeck, R. (2006), "A decision based analysis of compressed air usage patterns in automotive manufacturing," *Journal of Manufacturing Systems* 25, no. 4, 293–300.

45. Huijbregts, A. M. J., Hellweg, S., Frischknecht, R., Hungerbühler, K., and Hendriks, A. J. (2008), "Ecological footprint accounting life cycle assessment of products," *Ecological Economics* 64, 798–807.

46. Finnveden, G. (2008), "A world with CO_2-caps: Electricity production in consequential assessments," *International Journal of Life Cycle Assessment* 13, 365–367.

47. Eriksson, O., Finnveden, G., Ekvall, T., and Björklund, A. (2007), "Life cycle assessment of fuels for district heating: A comparison of waste incineration, biomass- and natural gas combustion," *Energy Policy* 35, 1346–1362.

48. Finnveden, G., Hauschild, M. Z., Ekvall, T., Guinee, J., Heijungs, R., Hellweg, S., Koehler, A., Pennington, D., and Suh, S. (2009), "Recent developments in Life Cycle Assessment," *Journal of Environmental Management* 91, 1–21.

49. Manuele, F. (2008), "Prevention through design: Addressing occupational risks in the design and redesign processes," *Professional Safety* 53, no. 10, 28–40.

50. Anderson, D. (2005), *Corporate Survival: The Critical Importance of Sustainability Risk Management* (New York: iUniverse, 2005), 1–15.

51. Boyle, C. A., "Achieving Sustainability," International Conference on Sustainability Engineering and Science (Auckland, July 7–9, 2004).

52. "Risk Assessment," http://epa.gov/riskassessment/basicinformation.htm#arisk, retrieved January 20, 2015.

53. Markowskia, A. S., and Mannanb, M. S. (2008), "Fuzzy risk matrix," *Journal of Hazardous Materials* 159, no. 1, 152–157.

54. "Environmental Assessments & Environmental Impact Statements," www.epa.gov/reg3esd1/nepa/eis.htm#eis, retrieved January 20, 2015.

55. "Environmental Assessment," http://ec.europa.eu/environment/eia/home.htm, retrieved January 20, 2015.

56. Akhavan, P., Elahi, B., and Jafari, M. (2014), "A new integrated knowledge model in supplier selection: The case of an Asian automotive supply chain," *Journal of Education, Business and Society: Contemporary Middle Eastern Issues* 7, no. 4, 333–368.

57. World Commission on Environment and Development, *Our Common Future* (New York: Oxford University Press, 1987), 5–30.

58. Palme, U., and Tillman, A. M. (2008), "Sustainable development indicators: How are they used in Swedish water utilities?" *Journal of Cleaner Production* 16, 1346–1357.

59. Khalili, N. R., and Duecker, S. (2013), "Application of multi-criteria decision analysis in design of sustainable environmental management system framework," *Journal of Cleaner Production* 47, 188–198.

60. "A Sustainability SWOT," www.wri.org/publication/sswot, retrieved January 23, 2014.

61. Stern, N., *The Economics of Climate Change: The Stern Review* (New York: Cambridge University Press, 2007), 140–162.

62. Pohekar, S. D., Ramachandran, M. (2004), "Application of multi-criteria decision making to sustainable energy planning—A review," *Renewable and Sustainable Energy Reviews* 8, 365–381.

63. Jaafar, I. H., Venkatachalam, A., Joshi, K., Ungureanu, A. C., De Silva, N., Dillon, J. O. W., Rouch, K. E., and Jawahir, I. S., "Product Design for Sustainability: A New Assessment Methodology and Case Studies," in *Handbook of Environmentally Conscious Mechanical Design* (New York: John Wiley & Sons, 2007), 25–65.

64. Jayal, A. D., Badurdeen, F., Dillon O. W., Jr., and Jawahir, I. S. (2010), "Sustainable manufacturing: Modeling and optimization challenges at the product, process and system levels," *CIRP Journal of Manufacturing Science and Technology* 2, 144–152.

65. Kreis, D., Howell, B., and O'Connell, L., "Early Stage Cost Analysis for Estimating Economic Impacts," Report No. KTC-06-11/SPR306-05-1F (Lexington: Kentucky Transportation Center, 2006).

66. Jiang, Y., Zhao, G., and Li, S., *An Economic Analysis Methodology for Project Evaluation and Programming*, SPR-3429, Report Number: FHWA/IN/JTRP-2013/17 (Joint Transportation Research Program, Indiana Department of Transportation and Purdue University, 2013)..

67. *A Guide to the Project Management Body of Knowledge (PMBOK® Guide)*, 5th ed. (Newtown Square, PA: Project Management Institute, 2013), 12–50.

68. Franchetti, M. 2013, "Economic and Operational Feasibility Analysis of Solid Waste Minimization Projects," Chap. 14 of *Integrated Waste Management*, Vol. 1, edited by Sunil Kumar (Rijeka, Croatia: InTech, 2013).

69. Kellerer, H., Pferschy, U., and Pisinger, D. *Knapsack Problems* (Berlin: Springer, 2004), 235–283.

70. CPLEX software, available at: www-03.ibm.com/software/products/en/ibmilogc pleoptistud, retrieved November, 2014.

71. Koc, A., Morton, D. P., Popova, E., Hess, S. M., Kee, E., and Richards, D. (2009), "Prioritizing project selection," *The Engineering Economist* 54, 267–297.

72. Lu, T., Gupta, A., Jaya, A. D., Badurdeen, F., Feng, S. C., Dillon, O. W., and Jawahir, I. S., "A Framework of Product and Process Metrics for Sustainable Manufacturing," in *Advances in Sustainable Manufacturing,* edited by G. Seliger, M. K. Khraisheh, and I. S. Jawahir (New York: Springer, 2011), 331–336.

73. Neugebauer, R., Wertheim, R., and Harzbecker, C., "Energy and Resources Efficiency in the Metal Cutting Industry," in *Advances in Sustainable Manufacturing*, edited by G. Seliger, M. K. Khraisheh, and I. S. Jawahir (New York: Springer, 2011), 247–257.

74. Wang, L., and Bi, Z. M., "Challenges for Better Sustainable Manufacturing," in *Advances in Sustainable and Competitive Manufacturing Systems*, edited by A. Azevedo (Stockholm: Springer, 2013), 1209–1221.

75. Elahi, B., Pakzad-Jafarabadi, Y., Etaati, L., and Seyedhosseini, S. M. (2011), "Optimization of supply chain planning with considering defective rates of products in each echelon," *Journal of Technology and Investment* 2, no. 3, 211–221.

76. Elahi, B., and Franchetti, M., "A New Optimization Model for Closed-Loop Supply Chain Networks, Technology Management Conference," 2014 IEEE International (Chicago, 2014), 1–9.

77. Elahi, B., Etaati, L., and Seyed-Hosseini, S. M. "An Integrated Modeling for Supplier Selection and Optimal Lot Sizing: A Case Study of Four-Echelon Supply Chain," Technology Management Conference, IEEE International (San Jose, 2011), 877–884.

78. Boyd, S., Kim, S. J., Vandenberghe, L., and Hassibi, A. (2007), "A tutorial on geometric programming," *Journal of Optimization and Engineering* 8, 67–127.

79. Seyedhosseini, S. M., Elahi, B., and Akhlaghy, A. (2011), "Proposing a non-linear mathematical model for order splitting in a supply chain with perishable products: Solving by genetic algorithm," *International Journal of Business Research* 11, no. 4, 104–111.

80. Mahdavi Mazdeh, M., Razavi, S. M., Hesamamiri, R., Zahedi, M. R., and Elahi, B. (2013), "An empirical investigation of entrepreneurship intensity in Iranian state universities," *Journal of Higher Education* 65, no. 2, 207–226.

81. Bertsimas, D., and Nohadani, O. (2009), "Robust optimization for unconstrained simulation-based problems," *Operations Research* 58, no. 1, 161–178.

82. Bertsimas, D., Brown, D. B., and Caramanis, C. (2011), "Theory and applications of robust optimization," *SIAM Review* 53, no. 3, 464–501.

83. Bertsimas, D., and Thiele, A. (2006), "A robust optimization approach to supply chain management," *Operations Research* 54, 150–168.

84. Ben-Tal, A., Ghaoui, L. E., and Nemirovski, A., *Robust Optimization* (Princeton, NJ: Princeton University Press, 2009), 1–26.

85. Jolliffe, I. T., *Principal Component Analysis* (New York: Springer, 2002), 10–28.

86. Jorgensen, F., Mathiessen, R., Nielsen, J., and Johansen, J., "Lean Maturity, Lean Sustainability," in *Advances in Production Management Systems*, Vol. 246, edited by J. Olhager and F. Persson (Boston: Springer, 2007), 371–378.

87. Shah, R., and Ward, P. T. (2003), "Lean manufacturing: Context, practice bundles, and performance," *Journal of Operations Management* 21, 129–149.

88. Yang, M. G., Hong, P., and Modi, S. B. (2011), "Impact of lean manufacturing and environmental management on business performance: An empirical study of manufacturing firms," *International Journal of Production Economics* 129, 251–261.

89. Cunha, V. P., Balkaya, I., Palacios, J., Rozenfeld, H., and Seliger, G., "Development of Technology Roadmap for Remanufacturing Oriented Production Equipment" in *Advances in Sustainable Manufacturing*, edited by G. Seliger, M. K. Khraisheh, and I. S. Jawahir (New York: Springer, 2011), 201–206.

90. Diaz-Elsayed, N., Jondral, A., Greinacher, S., Dornfeld, D., and Lanza, G. (2013), "Assessment of lean and green strategies by simulation of manufacturing systems in discrete production environments," *CIRP Annals—Manufacturing Technology* 62, 475–478.

91. Rivera, L., and Chen, F. F. (2007), "Measuring the impact of Lean tools on the cost–time investment of a product using cost–time profiles," *Journal of Robotics and Computer-Integrated Manufacturing* 23, 684–689.

Chapter 4

Sustainable Facility Management

Perspectives on the manufacturing footprint of global firms are broadening from the economic aspects to also comprise the environmental and social aspects. Considering such a situation, sustainable facility management activities have a significant influence over how buildings and facilities of a manufacturing system are used and thus are tasked to promote and fulfill the sustainability policies. Sustainability continues to have an effect on the design, construction, and operation of facilities universally. Sustainable facility management is the procedure of assimilating the people, place, and business of an organization or production/manufacturing system to such a degree that enhances economic, environmental, and social benefits of sustainability.

Sustainable facility management has a significant role in meeting strategic sustainability and environmental objectives and goals for the individual businesses or manufacturing systems in terms of providing them with a more efficient design or refurbishment result in efficient buildings and facility.[1, 2, 3]

As an initial action, in 2007 the International Facility Management Association (IFMA) emphasized implementing sustainability practices for various facilities of manufacturing systems. Facility managers can play fundamental roles in developing and fulfilling sustainability plans to reduce, reuse, and recycle waste and in working closely with end users to predict changes and conserve energy. The tasks can comprise assessing or monitoring the amount of energy used by the facilities they are supervising and adopting energy efficiency measures such as switching to efficient lighting equipment and adjusting heating and cooling and ventilation equipment to facility loads to decrease energy consumption. An adept facility manager is proficient in planning, managing/directing, and supporting the manufacturing system's commitment to the sustainability of the built and natural environments and the protection of the environment. Other competencies refer to developing and directing the defined sustainability programs regarding the facility, providing data to support facility evaluation in terms of sustainability, and evaluating and managing/overseeing the asset life cycle process. This is in accordance with the developing body of research in facilities management, which implies that energy management, waste management and recycling, transportation, carbon footprint, and environmental responsibility are the quintessential sustainability issues being tackled in manufacturing systems.

In agreement with the Chartered Institute of Building, in 2013 the built environment was incorporated into the sustainability agenda. In this agenda the built environment accounts for nearly 40% of limited natural resources consumed and 40% of waste and greenhouse gases generated. In fact, existing building stocks use as much as 45% of generated energy to produce power and heat. Alongside increasing utility and maintenance costs, coupled with rising legislative and regulatory requirements on energy use and carbon reduction, many organizations and manufacturing systems, committed to the sustainability agenda, have established sustainability policies as a central element of their corporate social responsibility (CSR).[1, 2, 3]

Moreover, sustainability is becoming an essential subject in determining the location of manufacturing facilities. Facility location decisions concern the geographical facilities' position in a particular structural entity, such as a manufacturing system. As such, they are strategic decisions related to the configuration of the manufacturing network. The literature on facility location can be categorized into two themes: factor assessment and mathematical approaches.

The factor assessment approach often has an emphasis on strategic issues in decision making, and it can be specified by four stages:

1. Setting up the critical success factors (CSFs) for a manufacturing system
2. Assessing alternatives for regional manufacturing configurations
3. Defining numerous potential sites
4. Ranking the most appropriate solutions

In sustainable facility location decisions, the objectives can be listed as follows:

1. Focusing on a cost-optimal distribution of facilities pertaining to the location of markets (final customers) and raw materials (suppliers)
2. Having more access to skills and knowledge regarding sustainability procedures and applying sustainable operations management
3. Concentrating on cost-effective energy options and identifying possible energy recovery management methods that can be adopted
4. Conducting plant capacity studies based on various scenarios and several sizes
5. Having estimations of pollutant emissions (such as greenhouse gas emissions) and evaluating the amount of generated waste and its composition
6. Considering sustainability factors toward manufacturing facility location

In terms of environmental evaluations, various sustainability factors such as ecosystem vitality (air pollution interrelated to the ecosystem, water quality linked with the ecosystem, biodiversity protection), environmental health (environmental burden of disease, air pollution related to humans, water quality related to humans), and environmental factors within production (material use, energy consumption, renewable resources, waste generation, waste treatment, waste disposal, recyclable material, energy, and waste) can be taken into account.[4], [5], [6] Building information modeling (BIM) is one of the useful new technologies that offer ways to facility managers to bridge the gap between the design and construction phase and the operation of buildings for manufacturing systems by making much more information available than existed in the past. The three components of this method are:

- Building: the whole life cycle of the buildings is taken into consideration (design/build/operations).
- Information: all information about the building and its life cycle is incorporated.
- Modeling: outlining and simulating the building, its delivery, and operations by means of integrated tools.

For example, an expert may discover proof of a leak in the facility. Instead of investigating the physical building, the expert may refer to the model and notice that the water valve is located in a doubtful location. He could also have in the model the specific valve size, manufacturer, part number, and any other data ever researched in the past, pending sufficient computing power.[7] Dynamic data related to the building, such as sensor measurements and control signals from the building systems, can also be integrated within BIM to support analysis of building operation and maintenance. BIM services facilitate the creation of models that serve as virtual representations of the construction processes by matching each phase with a frame-by-frame illustration. Using BIM can result in standardizing and streamlining the quality, mitigating risks, and detecting and removing conflicts and errors efficiently. Figure 4.1 shows different sources of information that are integrated with BIM in order to enhance the outcomes of modeling.

4.1. Facility Pollution Prevention Program

The facility pollution prevention outline was developed by the U.S. Environmental Protection Agency (EPA) in 1992 to assist small to medium-size production facilities to create broad-based multimedia pollution prevention (P2) programs in all enterprises and geographic locations. It explains how to recognize, assess, and implement opportunities for preventing pollution

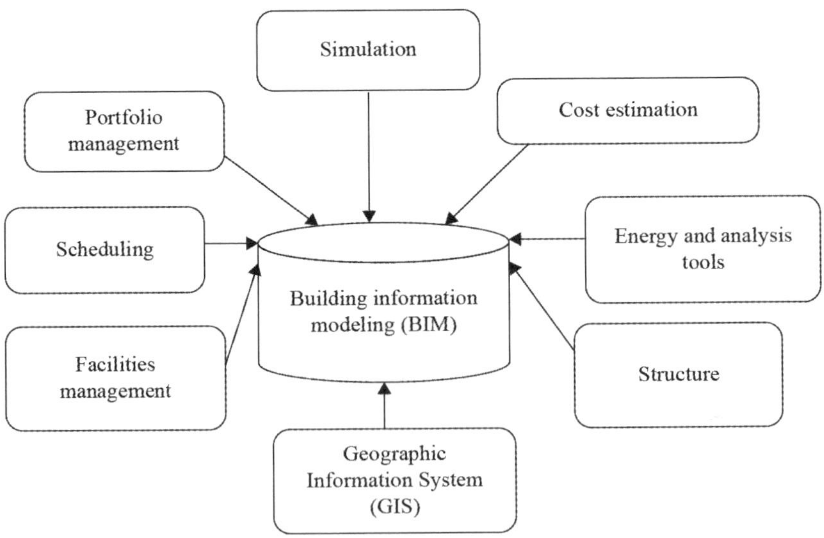

Figure 4.1. Collecting Different Sources of Information and Integration of Various Tools for Building Information Modeling

and how to encourage the continuing exploration of these opportunities.[8, 9, 10] As shown in Figure 4.2, the facility pollution prevention outline contains a series of successive phases. These phases can be categorized as follows:

1. Establishing the P2 program. (It must be reconsidered in accordance with the feedback of primary assessment.)
2. Organizing the P2 program.
3. Doing a preliminary assessment. (See "Note" below.)
4. Creating a program plan.
5. Performing a detailed assessment.
6. Defining P2 options and cost consideration.
7. Doing a feasibility study and a feasibility analysis.
8. Preparing an assessment report.
9. Implementing the plan.
10. Measuring the progress.
11. Maintaining the P2 program.

Note: To make a reliable environmental assessment, it is recommended that data be gathered on the environmental aspects of the related product, raw material, or essential parts of the process. These data would take the environmental effects into consideration not only in the production stage and product life cycle but also in the extraction and transportation of the raw materials and in treating waste. Throughout the environmental assessment, energy consumption should also be taken into consideration in the entire life cycle.

4.2. Sustainable Design and Layout

Sustainable design and site layout assist in reducing the traveling time and movement costs of plant, labor, and materials; activity interference throughout construction work; site accidents; and wastes. They assure that working on buildings and other construction locations is not hindered by the inattentive storage of materials on these sites. During the past five years, researchers have focused on different topics on various aspects of sustainable

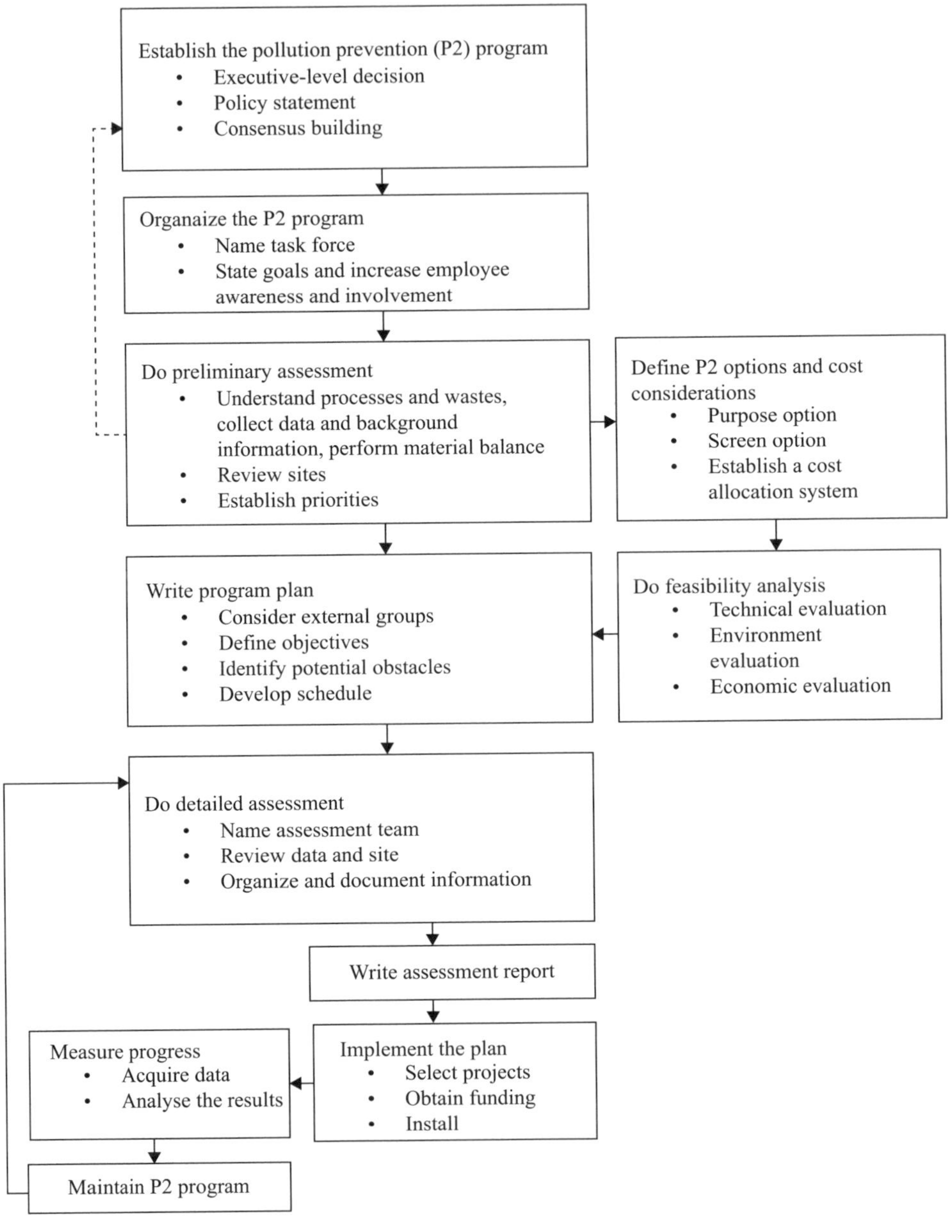

Figure 4.2. Facility Pollution Prevention (P2) Overview

design and site layout, such as space conflict, the design sequence of temporary facilities, the location of material and equipment, the intelligent equipment selection system, the intelligent materials routing system, multiconstraint site planning, the genetic algorithm for construction site layout, the genetic optimization of site layout, routing system for large vehicles, and Geographic Information System–based site layout planning system.[11] Industry researchers have also tried to distinguish opportunities to increase thermal energy efficiency of factories where possible and to minimize the energy intensity of operations.

A sample of sustainable design and site layout refers to energy-oriented design and layout planning for production facilities. A suitable design and layout planning can improve the energy efficiency of manufacturing processes. Regarding this subject, one method refers to the integration of energy efficiency into the planning process and developing a planning system for the detailed projection of a production system's energy consumption (with the concept of energy blocks). With the help of such an approach, a time-based forecast of the amount of energy required by each machine and the entire manufacturing system will be available.

Once setting up energy efficiency goals, the strategic aspect of the layout planning of a manufacturing area in a facility (e.g., production area or assembly area) should be associated with the energy flow throughout the production process (e.g., the input and output of energy in every production process or cell). While working on a layout plan, an expert should consider the energy consumption profile of the manufacturing system to be taken into account. It denotes that an assessment of the various potential alternatives for designing a layout based on its energy efficiency should take place. This assessment is only feasible once the data in relation to the consumption behavior of the individual equipment are accessible. Throughout the design stage of the layout, the planner can assess the different options for each of production systems and define the optimal manufacturing system configuration by using an energy database. Energy flow is an integral component in layout planning. By applying this method, it is achievable for the manufacturing systems to attain an energy-efficient status during layout planning. This method begins with a major consideration that each of the production facilities has a related energy demand, which entails environment requirements regarding energy usage and a variety of energy consumptions (for example, electricity and compressed air), and an energy output (for instance, the thermal exhausting and the energy bounded on products). Energy demand and energy output in each of the manufacturing processes are the centers for energy-efficient layout planning, when it can be designed to achieve an energy equivalence status among different facilities. Based on energy efficiency standpoint, a sustainable facility layout refers to minimizing energy loss throughout the entire production processes. Therefore, the reuse or an appropriate organization of energy outputs from one production facility to another directly or by means of special equipment (for instance, energy storage) will be important for an energy efficiency enhancement in a manufacturing system.

Once energy consumption and losses, which can be classified in manufacturing, assembly, and transportation, throughout the entire production processes in one or multiple facilities are understood and evaluated, a comprehensive database for the used energy can be set up. Such a wide-ranging database involves an energy input database (electricity, heat, hydraulic, pneumatics, etc.), energy storage database (energy harvesting, energy storage methods, energy conversion efficiency, etc.), and energy output database (heat losses, vibrations, etc.). Moreover, the exploitation analysis of energy loss contributes to the energy-oriented layout planning (for instance, determining various energy outputs related to one facility that can be applied to others and defining the proper method regarding how these data can be utilized). Hence, an energy-based facility layout can be designed in association with energy analysis.

Furthermore, a proper simulation can be applied to achieve a sustainable, energy-oriented layout plan for manufacturing facilities by means of accessing the energy database and monitoring energy flows. Simulation approaches are effective tools to model the energy profiles related to each individual process within a manufacturing/production system. By combining process modeling with production metrics, energy metrics, and current energy price, an optimal schedule can be achieved.

4.3. Leadership in Energy and Environmental Design

Leadership in Energy and Environmental Design (LEED) is a set of rating systems and a third-party ecology-oriented certification program for the design, construction, operation, and maintenance of green buildings advanced by the U.S. Green Building Council (USGBC). LEED is planned to assist building owners, managers, engineers, and operators who are environmentally responsible for applying resources efficiently. This rating system applies strategies used for improving performances across all the metrics that matter most: energy savings, water efficiency, CO_2 emissions reduction, enhanced indoor environmental quality, and stewardship of resources and sensitivity to their impacts. LEED identifies the best-in-class building strategies and practices (for example, energy-efficient lighting, low-flow plumbing fixtures, and collection of water). To receive a LEED certification, building projects meet the prerequisites and earn points to attain various levels of certification. Prerequisites and credits change for each of selected rating systems, and a team of decision makers selects the best fit for the related project. Recycled construction materials and energy-efficient appliances also affect the points in the rating system.

The LEED rating system presents four certification levels for new constructions and facility sites (Certified, Silver, Gold, and Platinum) that are compatible with the number of credits accumulated in five green design classes: sustainable sites, water efficiency, energy and atmosphere, materials and resources, and indoor environmental quality. Credits are split into individual points. LEED certification levels are as follows, and a building needs at least 40 points for certification in LEED for new construction, version 3.0:

- Certified, 40–49 Points
- Silver, 50–59 Points
- Gold, 60–79 Points
- Platinum, 80–110 Points

Support for green buildings has fostered rapidly each year over the past five years by the Department of Energy (DOE), the EPA, and other agencies. Many cities and states either offer tax credits or grants for green buildings or require green building certification for public buildings. Considering such a condition, setting a clear environmental target, allocating a clear and adequate budget, engineering for life cycle value, and hiring LEED-accredited professionals can help manufacturing systems be approved for LEED certification. Moreover, the Alternative Compliance Paths (ACPs) supplements are complementary to the 2009 edition of the *LEED Reference Guide for Green Building Design and Construction*. The ACPs integrated with LEED credits provide additional alternatives or methodologies that address the unique circumstances and facilitate advancements in science and technology. The ACPs allow LEED to be more flexible and practical for a broader range of projects.[12, 13].

4.4. Green Building and Infrastructure

In response to the global environment protection and sustainability movements, the construction industry has adopted the "green building" concept with reasonable success. In the green building industry and infrastructure, green construction cost controls and low-carbon construction procedures are taken into consideration as the chief obstacles encountered. According to CSR policy, controlling the carbon emissions of green building projects contributes to gaining the accurate building cost information and making reductions in the environmental impact of these projects. To attain environmentally sustainable development, construction companies and systems should focus on environmental and social costs such as CO_2 emission costs throughout the whole life cycle of green building assessment. Low-carbon construction procedures for green building projects can improve building durability and give rise to less construction facility waste (e.g., prefabrication of some building components). Four main stages in the life cycle of green building can be classified as follows:

1. **Green design stage.** Green design strategies promote sustainability by taking occupant health and environmental resource depletion into consideration throughout the construction. This stage entails green design consultant fees (e.g., construction's site selection and renewable energy source approval).

2. **Construction stage.** Unlike in the traditional construction process, in this stage, an environmental impact assessment is involved in the green build ing construction process. This stage covers the transportation of building materials and low-carbon construction procedures; it entails environment loads stemming from the use of heavy equipment or machines at the construction site.[14]

3. **Operation stage.** Energy consumption throughout the operation stage is classified into operational uses (electricity, heating, and ventilating) and maintenance and building refurbishing (repair) activities.[15]

4. **Decommissioning stage.** This stage entails the dismantling of the building structure shell and the final disposal activities: landfill use, waste recycling, and reuse.

Significant features that are related to the high performance green building include high energy efficiency (e.g., by the use of more efficient insulation systems, air barriers, seamless monolithic roofing systems, etc.), occupant comfort, material durability, and increased property values. Considering these features, green building practice is becoming institutionalized by various organizations, such as the U.S. Green Building Council through its LEED rating systems.

Based on the DOE, buildings require a substantial amount of energy; the total is about 40 quadrillion Btu (quad) per year, which is equal to 4.22×10^{19} joules, with commercial buildings accounting for more than 40% of the entire energy use. A quad is a unit of energy equal to 1.055×10^{18} joules in the International System of Units. Along with energy consumption, buildings account for more than 40% of all CO_2 emissions in the United States, mainly because of their dependence on coal-fired electrical plants for electricity supply. These two factors, coupled with the growing cost of operating commercial and industrial buildings, make energy efficiency a key driver in sustainable design and construction in addition to the building operations, as shown in Figure 4.3.

4.4.1. Green Building Cost Assessment

From the perspective of construction companies and systems, it is vital to appraise the costs of the related construction process tasks for each green building project to be able to evaluate the total cost. The activity-based costing (ABC) concept can be applied to assess the green building costs. The ABC analysis is classified into a two-stage procedure that allots resource costs to cost objects. Applying such a method, green building projects are considered as cost objects. In the first phase, resource costs are allocated to different activities; applying resource drivers, the factors are selected to estimate the consumption of resources utilized in construction activities. Each type of resource traced to a building activity becomes one cost element within an activity cost pool. Hence, an activity cost pool gives the total costs related to a specific activity. An activity center is composed of the associated activities, usually classified based on function or process.

In the second phase, the costs in each activity cost pool are allocated to cost objects by a suitable activity driver applied to compute the consumption of activities by the cost objects. For this reason, decision makers can choose appropriate cost drivers that count on measurable objectives, costs of measurement, and the intensity of correlation.

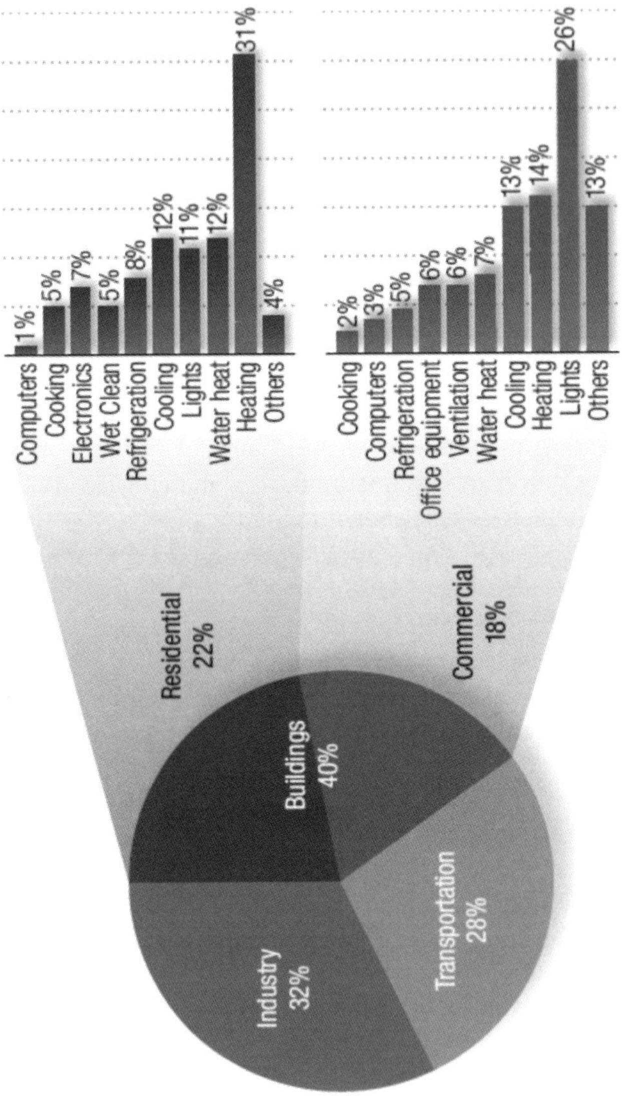

Figure 4.3. Profile of Energy Use in Different Sectors

Source: U.S. Energy Information Administration, 2012.

Costs that are integrated with the whole life cycle of a green building project entail the costs of direct materials, direct labor, machines, land acquisition, activity costs, CO_2 emissions, and other costs that can be looked at as unchangeable fixed costs. In a green building project, costs are classified into direct and indirect construction costs. Direct costs are categorized into direct material costs, direct labor costs, direct machine costs, and land acquisition costs. Indirect costs are categorized into four activity levels containing: unit, batch, environment, and project. Unit-level activities are related to the activities that should be carried out for each unit of cost object. Batch-level activities are associated with each batch of cost objects, for example, material transportation for green building. Environment-level activities entail activities that are attached to the environmental energy of a cost object, for example, energy management. Project-level activities are conducted as needed to support all units of a cost object, for example, design planning for green building.[16]

4.4.2. Indoor Environmental Quality

Indoor environmental quality (IEQ) is one of the important topics in green building, especially in large buildings where the occupants' health and workplace productivity are essential. There are various subjects in the area of IEQ, such as indoor air quality, comfort, effective operation, control of thermal and lighting systems, and adequate occupant exposure to daylight. Managing and monitoring these variables are vital in green buildings. Various conducted research demonstrates a direct correlation between indoor pollutants and the accurateness of specific tasks; in addition, they display a particular correlation between the distinguished indoor pollutants and the performance of work tasks. Related to the study of IEQ, an outstanding research and statistical analysis was performed by the Lawrence Berkeley National Laboratory (LBNL) in 2011. This analysis demonstrated that when ventilation rates in office buildings are doubled, a decrease in short-term absences for more than one-third of the occupants was seen. In the LBNL research, buildings with above-average ventilation rates showed considerably fewer symptoms of sick building syndrome in workplace occupants, with reductions of 10% to 80%.

The LBNL also illustrated that workplace performance drops off significantly in terms of speed and accuracy when interior temperatures increase above, or decrease below, 21.7°C (71°F). To cope with these concerns, green building criteria necessitate that heating, ventilating, and air-conditioning (HVAC) systems and the building envelope (insulation and air tightness) be designed in a way that allow for optimal thermal controls. ANSI/ASHRAE Standard 55-2010 entails the factors of humidity, air temperature and speed, and radiant temperature. Essentially, in order to effectively control ventilation rates, indoor temperature, and humidity, the building envelope should be tightly sealed to avoid uncontrolled air infiltration.[17]

4.4.3. Insulation Systems

Sustainable building and facility construction starts with reducing the heating and cooling loads as much as possible, and insulation is the key to energy conservation, a cornerstone of green building. Well-insulated buildings and facility sites not only save energy, thus lowering operating costs, but also keep people more comfortable. Selection of insulation depends on a large number of factors, such as availability of the space for insulation, climatic conditions, availability of funds, and potential savings. There are a number of different insulation options available to businesses and manufacturing systems; some examples follow:

- Blanket rolls
- Concrete block insulation
- Foam board insulation
- Loose-fill insulation
- Reflective insulation
- Spray foams
- Structural insulated panels

Proper selection of the insulation can make a significant difference in terms of quality and costs. Structural insulated panels and insulated concrete forms have shown increased energy savings of up to 25% over competing materials.[18]

- **Blanket rolls.** Insulation materials are fiberglass, mineral wool (in the form of rock or slag), plastic fibers, and natural fibers. They can be applied to unfinished walls, entailing foundation walls, floors, and ceilings. They fit between studs, joists, and beams. Its typical R-value per inch of thickness belongs in the range of {3.7, 4.7}. The R-value is a measure of thermal resistance used for a building or facility's construction. In uniform conditions, it is the ratio of the temperature difference (ΔT) across an insulator and the heat flux (heat transfer per unit area per unit of time, $\dot{Q}_A$) through it. It can be stated by the following formula:

$$R = \frac{\Delta \mathrm{T}}{\dot{Q}_A} \tag{1}$$

Thermal resistance changes with temperature; however, it is common practice in construction to consider it as a constant value. The R-value depends on a solid material's resistance to conductive heat transfer. The R-value is explained as the thickness of the material normalized to the thermal conductivity (the higher the value, the more improvement in the building insulation's theoretical efficiency).

- **Concrete block insulation.** Insulation materials are foam beads or liquid foam from polystyrene, polyurethane, or vermiculite or perlite pellets. It can be used for unfinished walls consisting of foundation walls for new construction or main renovations. Its typical R-value per inch of thickness belongs to the range of {1.1, 2}.

- **Foam board or rigid foam.** Insulation materials are polystyrene and polyurethane. It can be applied to unfinished walls consisting of foundation walls, floors and ceilings, and unvented low-slope roofs. The typical R-value per inch of thickness for polystyrene belongs to the range of {3.8, 4.4}and for polyurethane{5.6, 8}.

- **Insulated concrete forms.** Insulation materials are foam boards or foam blocks. They are applicable for unfinished walls, including foundation walls and new construction. Their typical R-value per inch of thickness is 17.

- **Loose-fill.** Insulation materials are cellulose, fiberglass, and mineral (rock or slag) wool. It can be applied to enclosed, existing wall or open new wall cavities; unfinished attic floors; or hard-to-reach places. The typical R-value per inch of thickness for cellulose belongs to the range of {3.2, 3.8}, for fiberglass {2.2, 2.7}, and for rock wool {3, 3.3}.

- **Reflective system.** Insulation materials are foil-faced kraft paper, plastic film, polyethylene bubbles, or cardboard. It can be used for unfinished walls, ceilings, and floors. Its typical R-value per inch of thickness belongs to the range of {4, 15}.

- **Rigid fibrous or fiber insulation.** Insulation materials are fiberglass and mineral (rock or slag) wool. It can be applied to ducts in unconditioned spaces and other places having need of insulation that can resist high temperatures. Its typical R-value per inch of thickness is 4.
- **Sprayed foam and foamed-in-place.** Insulation materials are phenolic and polyurethane. They can be used for enclosed walls, cavities, and un finished attic floors. Typical R-value for low-density 5.5-inch (about 14-centimeter) sprayed polyurethane foam is 19.8, and for high-density 4-inch (about 10-centimeter) sprayed polyurethane foam, it is 27.
- **Structural insulated panels.** Insulation materials are foam board or liquid foam insulation core, and straw core insulation. They are used for unfinished walls, ceilings, floors, and roofs for new construction. The typical R-value per inch of thickness for polystyrene is 4, and for polyurethane it belongs to the range of {6, 7}.

Moreover, a window's surface can be one of the major sources of heat loss. The heat loss may be estimated by knowing the area of the window, temperature difference (average daily value of the year x 365), and U-value (thermal transmittance coefficient) of the window. The U-values for new windows are available from the manufacturer. Typically, U-values of new windows are at or below 0.35. The U-value of a window is the rate of loss of heat in watts per square meter of that window per degree of centigrade temperature difference across the window. This value can also be evaluated by other measurement systems, such as U.S. customary units. The lower the U-value, the greater is the window's insulating value. Estimation of the payback period for window replacement can be performed by calculating saved wattage and converting it into an annual energy savings.[18]

References

1. International Facility Management Association, *Facility Management Forecast 2007: Exploring the Current Trends and Future Outlook for Facility Management Professionals* (Houston: IFMA, 2007), 2–17.

2. Elmualim, A., Valle, R., and Kwawu, W. (2012), "Discerning policy and drivers for sustainable facilities management practice," *International Journal of Sustainable Built Environment* 1, 16–25.

3. Chartered Institute of Building, *Submission to the All Party Parliamentary Group for Excellence in the Built Environment on Inquiry into Sustainable Construction and the Green Deal* (Ascot, Berkshire, England: CIOB, 2013), 1–6.

4. Reuter, C., Foerstl, K. A. I., Hartmann, E. V. I., and Blome, C. (2010), "Sustainable global supplier management: The role of dynamic capabilities in achieving competitive advantage," *Journal of Supply Chain Management* 46, no. 2, 45–63.

5. Chen, L., Olhager, J., and Tang, O. (2014), "Manufacturing facility location and sustainability: A literature review and research agenda," *International Journal of Production Economics* 149, 154–163.

6. Cucchiella, F., Adamo, D. I., and Gastaldi, M. (2014), "Sustainable management of waste-to-energy facilities," *Renewable and Sustainable Energy Reviews* 33, 719–728.

7. Fernanda, L., and Burcu, A. (2012), "Formalized representation for supporting automated identification of critical assets in facilities during emergencies triggered by failures in building systems," *Journal of Computing in Civil Engineering* 26, no. 4, 519–529.

8. U.S. Environmental Protection Agency, *Facility Pollution Prevention Guide*, EPA/600/R-92/088 (Washington, DC: Office of Research and Development, 1992), 33–75

9. Ohio EPA, *Ohio Pollution Prevention and Waste Minimization Planning Guidance Manual* (Columbus, OH: Office of Pollution Prevention, 1993), chapters 6–7.

10. Hossain, K. A., Khan, F. I., and Hawboldt, K. (2008), "Sustainable development of process facilities: State-of-the-art review of pollution prevention frameworks," *Journal of Hazardous Materials* 150, 4–20.

11. Abdul-Rahman, H., Wang, C., and Eng, K. S. (2011), "Repertory grid technique in the development of Tacit-based Decision Support System (TDSS) for sustainable site layout planning," *Journal of Automation in Construction* 20, 818–829.

12. *LEED Reference Guide for Green Building Design and Construction with Global Alternative Compliance Paths,* rev. ed. (Washington, DC: U.S. Green Building Council, October 2014), 53–70.

13. U.S. Green Building Council, www.usgbc.org/leed, retrieved January 2015.

14. Li, X., Zhu, Y., and Zhang, Z. (2010), "An LCA-based environmental impact assessment model for construction processes," *Journal of Building and Environment* 45, no. 3, 766–775.

15. Blengini, G. A., and Carlo, T. D. (2010), "The changing role of life cycle phases, subsystems and materials in the LCA of low energy buildings," *Journal of Energy and Buildings* 42, no. 6, 869–880.

16. Tsai, W. H., Yang, C. H., Chang, J. C., and Lee, H. L. (2014), "An Activity-Based Costing decision model for life cycle assessment in green building projects," *European Journal of Operational Research* 238, 607–619.

17. "Indoor Environmental Quality," Environmental Energy Technologies Division (EETD), Lawrence Berkeley National Laboratory, http://eetd.lbl.gov/research-development/indoor-environmental-quality, retrieved January 2015.

18. Spivak, A., Kumar, A., and Franchetti, M., *Energy Assessments for Industrial Complexes* (Oak Park, OH: Bentham Science, 2013), 12–33.

Chapter 5

Environmental Resource Management and Pollution Prevention

Environmental resource management refers to the management of the interaction and impact of human activities on the environment. It is more than the management of the environment itself; it aims to ensure that environmental resources are utilized in an ethical, efficient, and economical way so that the environment is protected and maintained for future generations. Its purpose is to identify and manage factors affected by conflicts that arise from meeting the needs of society and, at the same time, protecting the environment from substantial degradation. It is thus linked to the concepts and practices of sustainability and environmental protection.

Pollution prevention refers to the elimination and minimization of pollutants and waste generation at the source, rather than reliance on management and treatment after they are created. In the United States, the Congress, the Environmental Protection Agency (EPA), and environmental professionals came to the conclusion in the 1980s that a new industrial waste management philosophy was needed if the ever-expanding industrial pollution and resource depletion problems were to be solved. Indiscriminate use of virgin resources in manufacturing and subsequent end-of-pipe treatment of resulting wastes would not provide long-term resource sustainability and environmental quality. As a result, a new paradigm was developed that emphasized minimizing the use of harmful or over-exploited resources and eliminating or minimizing waste production at the source in the industry's production area. This philosophy became known by many names, including *waste minimization, source reduction, green engineering,* and *sustainable engineering,* but the name that is most often associated with it is *pollution prevention*.

The EPA defines *pollution prevention* as the use of materials, processes, or practices that reduce or eliminate the creation of pollutants or wastes at the source.[1] It includes practices that reduce the use of hazardous materials, energy, and water or other resources and practices that protect the natural resources through conservation or more efficient use.

Source Reduction

I. Product Changes

- Design for less environmental impact
- Increase product life

II. Process Changes

A. Input Material Changes
- Material purification
- Substitution of less toxic materials

B. Technology Changes
- Layout changes
- Increased automation
- Improved operating conditions
- Improved equipment
- New technology

C. Improved Operating Practices
- Operating and maintenance procedures
- Management practices
- Stream segregation
- Material handling improvements
- Production scheduling
- Inventory control
- Training
- Waste segregation

Figure 5.1. Typical Source Reduction Methods
Source: Adapted from the U.S. Environmental Protection Agency.

Thus pollution prevention includes both the modification of industrial processes to minimize the production of wastes and the implementation of sustainability concepts to conserve valuable resources. Pollution prevention activities range from product changes to process changes to changes in methods of operation. This wide variety of activities is depicted in Figure 5.1.

5.1. Regulatory Background

Through the mid-1960s and early 1970s, there was a major outcry by the U.S. public and scientific community to clean up the environment and prevent further degradation. This led to the passage of several pieces of important regulation by Congress pertaining to clean air, clean water, hazardous materials and wastes, and pollution prevention. This section briefly discusses some of the key environmental laws and regulations.

5.1.1. Laws Pertaining to Clean Air

The first laws established in the United States to protect and enhance the quality of the nation's air resources were passed in 1955 (Air Pollution Control Act).[2] These have since undergone several major changes and amendments, notably in 1967, 1970, and 1977, all of which were to the Clean Air Act (CAA). The last major change in the laws took place in 1990 and was known as the Clean Air Act Amendments of 1990. This included revisions that were designed to improve its effectiveness and to target newly recognized air pollution problems such as acid rain and damage to the stratospheric ozone layer.

To protect public health and welfare nationwide, the CAA requires the EPA to establish national ambient air quality standards for certain common and widespread pollutants based on the latest science. The EPA has set air quality standards for six common "criteria pollutants": particulate matter (also known as particle pollution), ozone, sulfur dioxide, nitrogen dioxide, carbon monoxide, and lead. States are required to adopt enforceable plans to achieve and maintain air quality meeting the air quality standards. State plans also must control emissions that drift across state lines and harm air quality in downwind states. Other key provisions are designed to minimize pollution increases from growing numbers of motor vehicles, and from new or expanded industrial plants. The law calls for new stationary sources (e.g., power plants and factories) to use the best available technology, and it allows less stringent standards for existing sources.

The act also contains specific provisions to address hazardous or toxic air pollutants that pose such health risks as cancer or environmental threats such as bioaccumulation of heavy metals; acid rain that damages aquatic life, forests, and property; chemical emissions that deplete the stratospheric ozone layer, which protects us from skin cancer and eye damage; and regional haze that impairs visibility in national parks and other recreational areas. In addition to creating programs to solve identified pollution problems, Congress drafted the act with general authorities that can be used to address pollution problems that emerge over time, such as greenhouse gases which cause climate change.

5.1.2. Laws Pertaining to Clean Water

The federal Water Pollution Control Act was originally passed in the United States in 1948 and was aimed to protect the nation's surface waters, such as lakes, rivers, and coastal areas.[3] This was totally rewritten and expanded in 1972 and became known as the Clean Water Act (CWA). It has since been revised several times, with major revisions in 1977, 1981, and 1987.

The Clean Water Act established the basic structure for regulating discharges of pollutants into U.S. waters and regulating quality standards for surface waters. With the objective of improving and maintaining water quality by restoring the physical, chemical, and biological integrity of the nation's water bodies, the CWA has five main elements:

1. A system of minimum national effluent standards for each industry
2. Water quality standards
3. A discharge permit program that translates these standards into enforceable limits
4. Provisions for special waste such as toxic chemicals and oil spills
5. A revolving construction program for publicly owned treatment works

Under the CWA, the EPA has implemented pollution control programs such as setting wastewater standards for industry. The CWA made it unlawful to discharge any pollutant from a point source into navigable waters, unless a permit was obtained. The EPA's National Pollutant Discharge Elimination System (NPDES) permit program controls discharges. Point sources are discrete conveyances such as pipes or manmade ditches. Individual homes that are connected to a municipal system, use a septic system, or do not have a surface discharge do not need an NPDES permit; however, industrial, municipal, and other facilities must obtain permits if their discharges go directly to surface waters.

5.1.3. Laws Pertaining to Hazardous Materials and Wastes

The Solid Waste Disposal Act of 1965 was one of the first comprehensive regulations in the United States to address solid waste management. This law, however, primarily dealt with the disposal of nonhazardous waste, and in 1976, Congress passed the Resource Conservation and Recovery Act (RCRA), which completely rewrote the previous legislation and provided the framework for national programs to achieve environmentally sound management of both hazardous and nonhazardous wastes. The hazardous waste program under the act established a permit program to manage potentially hazardous waste from its point of origin to its point of ultimate disposal. The solid waste program was aimed at developing comprehensive plans at the state level to manage nonhazardous industrial and municipal solid waste and to set criteria for municipal solid waste landfills and other solid waste disposal facilities. The underground storage tank program was set up to regulate underground storage tanks containing hazardous substances and petroleum products.

The RCRA was amended and expanded in scope in 1984 with the enactment of the Hazardous and Solid Waste Amendments (HSWA). A major component of the HSWA was legislation directed at minimizing contamination of groundwater by hazardous waste. Restriction was placed on land disposal of hazardous waste unless it could be demonstrated that there would be no migration of hazardous constituents from the disposal unit.

The Comprehensive Environmental Response, Compensation, and Liability Act (CERCLA), commonly known as Superfund, was enacted in 1980. This law created a tax on the chemical and petroleum industries and provided broad federal authority to respond directly to releases or threatened releases of hazardous substances that may endanger public health or the environment. The law authorizes short-term removal actions to address releases or threatened releases requiring prompt response, as well as long-term remedial response actions that permanently and significantly reduce the dangers associated with releases or threats of releases of hazardous substances. CERCLA also enabled the revision of the National Contingency Plan, which provided the guidelines and procedures needed to respond to releases and threatened releases of hazardous substances, pollutants, or contaminants. CERCLA was amended by the Superfund Amendments and Reauthorization Act in 1986.

5.1.4. Laws Pertaining to Pollution Prevention

The Pollution Prevention Act (PPA) of 1990 was aimed at establishing a completely different pollution management hierarchy system. All prior regulations on waste management dealt with end-of-pipe requirements, with the assumptions that creation of pollutants is inevitable and that the waste must be treated after it is produced. The goal of the PPA was to shift the nation's waste strategy from the control of waste after its generation to the reduction of waste at its source.[1] Pollution prevention was declared to be the nation's primary pollution control strategy. With this as the goal, the regulation decreed that pollution should be prevented or reduced at the source whenever feasible. Recognizing that not all waste can be prevented, it declared that pollution that cannot be prevented should be recycled if possible. Treatment and disposal in an environmentally safe manner should

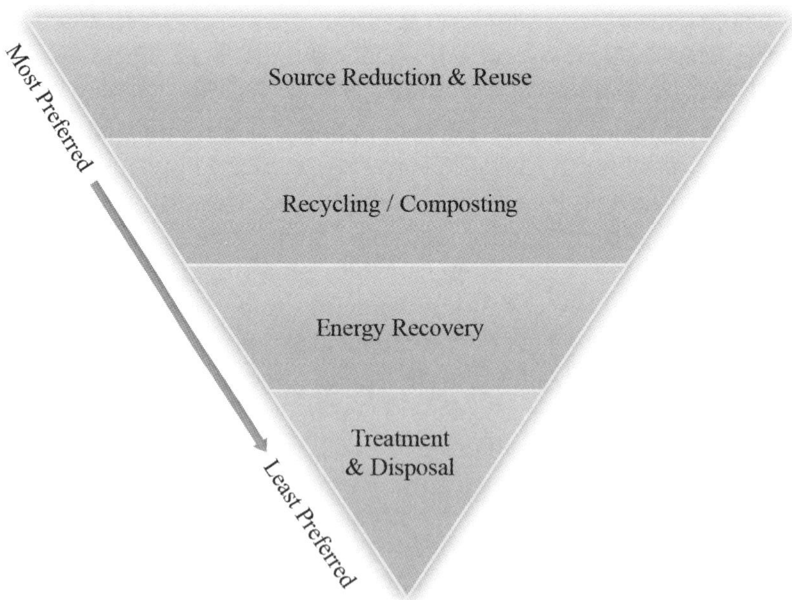

Figure 5.2. Waste Management Hierarchy
Source: U.S. Environmental Protection Agency.

be the last resort. The preferred waste management hierarchy is depicted in Figure 5.2.

The PPA authorized matching grants for states to establish technical assistance programs for businesses, and the EPA was directed to establish a Source Reduction Clearinghouse to compile information on source reduction and make it available to the public and the grant program. The act imposed new reporting requirements on industry; businesses were required to annually file a toxic chemical source reduction and recycling report with the EPA.

5.2. Benefits of Pollution Prevention

The purpose of all pollution prevention initiatives is to achieve measurable results that can be tied into the original goal. Results and benefits are often critical in determining the feasibility or acceptance of an initiative, leading to successful implementation. These benefits are also the key selling points used when promoting pollution prevention to stakeholders and decision makers. Ideally, any business would like a situation where multiple benefits could be realized from a particular pollution prevention initiative. Benefits can be broadly classified into four major areas:

- Environmental
- Economic
- Corporate image
- Social and personal

Key environmental benefits of pollution prevention and waste minimization include conservation of natural resources (minerals, water, energy, land, and forests), healthier environment and protection from further or irreversible degradation, and conservation of habitats. Most pollution prevention activities are initially begun because of regulatory pressure, cost savings associated with minimizing waste generation, or to improve corporate image; in recent years, however, many businesses are increasingly moving beyond these goals to pursue pollution prevention on a more moral basis. Pollution management decisions are being based on environmental ethics.

Economic benefits are often key selling points when promoting environmentally conscious initiatives to businesses. Economic incentives can be significant, depending on the type of industry, the nature of the processes, and the competitive business environment. Cost savings can include reduced waste treatment and disposal costs; reduced raw material costs; reduced transportation costs; reduced compliance costs for permits, monitoring, and enforcement; reduced production costs through better management and income derived through sale or reuse of waste; and reduced energy costs.

5.3. Water Use and Management

Water resources are sources of water that are used for agricultural, industrial, domestic, recreational, and other human activities. The majority of human uses require freshwater. With only about 3% of the total water on earth being freshwater and with over two-thirds of this being frozen in glaciers and polar ice caps, water resources are very unevenly distributed throughout the world. Most of the unfrozen freshwater is found as groundwater, with only a small fraction found aboveground as surface water and in the air. Freshwater is a renewable resource; however, long-term human activities have resulted in steady depletion of groundwater supplies around the world and natural water renewal is increasingly being unable to balance human usage.

5.3.1. Water Use in the United States

According to data from the U.S. Geological Survey, about 355 billion gallons per day (Bgal/d) of water was withdrawn for use in the United States in 2010.[4] About 6% of the total was for industrial use, withdrawn through public supply systems and industrial self-supply systems. Other uses include about 45% for thermoelectric power generation, 33% for irrigation, 7% for domestic use, 5% for commercial and similar use, 3% for livestock operation and aquaculture, and 1% for mining operation. Total water use in 2010 declined by about 13% from the level of 400,000 Bgal/d in 2005. Total public supply withdrawals for domestic, industrial, and commercial use in 2010 declined by 5% from 2005, and self-supplied industrial use was 12% less than in 2005. Substantial reductions in use in 2010 also occurred for thermoelectric power generation, irrigation and livestock operation; only mining operation and aquaculture underwent an increase in water use from 2005 levels.

5.3.2. Industrial Water Use

One major component of the resources used in industrial processes—water—is often overlooked. Water is thought of as an essentially "free" resource because of its relatively low cost; however, it is not free, and when used in large quantities, the cost can add up. It is treated potable water that is used in most industrial processes, and this is a valuable resource that should not be wasted. Another significant factor discouraging excessive water use is the cost of treating any resulting contaminated wastewater. Wastewater treatment is expensive, and the cost increases as the volume of water to be treated increases. If the water becomes contaminated with hazardous materials through industrial processes, the cost can become prohibitive. Thus, resource conservation aside, it is essential that water use be minimized to avoid costs associated with its treatment.

Many industries such as oil refineries, semiconductor manufacturers, and facilities that process fruits, vegetables, and meat require large amounts of water in their operation. Industry uses water for a variety of purposes, including processing needs (e.g., fabrication; dilution: as a solvent; material transport; and incorporation into a product), cooling, cleaning and sanitation needs within the facility, landscaping, in kitchens and restrooms, and for laundry. The majority of water used by an industrial facility is often used for processing, although cooling, most notably for oil refineries, high tech, and food processing facilities, also consumes a lot of water.

5.3.3. Common Costs Associated with Water Use

Costs associated with water use include more than the direct costs paid for water supply; they include the costs of water as it travels through processes and operations, often consuming energy in heating or cooling the water for process needs, and also the costs associated with wastewater treatment and disposal. Full costs associated with water use often vastly outweigh the direct costs. The direct cost of purchasing water from a utility may not seem significant enough alone to merit the effort to reduce water use, but when the full cost of water is assessed, the financial savings can be substantial. The following lists include the common costs associated with water use:[5]

Raw Material Costs

- Water purchased from utilities; marginal costs of purchasing additional water versus costs of conservation

- Cost of water treatment, filtering, and softening before use
- Cost of chemicals needed to treat and manage water

Energy Costs

- Cost of energy to heat water
- Cost of energy to pump water from its source or within the facility itself
- Energy and labor costs for operating and maintaining water-using equipment

Pollution Control Costs

- Wastewater and storm water service rates, including surcharges
- Total cost of treating wastewater for disposal, including labor, energy, chemicals, equipment, and residual disposal
- Marginal cost of increasing effluent treatment capacity when water demand increases

Regulatory Compliance Costs

- Labor costs for regulatory compliance activities such as completing permit applications, monitoring compliance, and reporting wastewater discharges to regulatory agencies

5.3.4. Water Use Management Techniques

In recent decades, management strategies and methods have been well developed for reducing the amount of water required for industrial operations without compromising quality. Simple techniques to foster an understanding of where water goes in a facility and where it can be saved include the following:

- Water use survey/facility walk-through
- Developing a water balance
- Water waste and value stream mapping
- Root cause analysis

While all of these techniques evaluate aspects of facility water use and can support water usage reduction efforts, each has a different level of detail, time, and staff investment requirement—ranging from a facility walk to a multiday value stream mapping event. It is necessary to consider the level of effort desired before getting started to conduct the most effective analysis for a facility's needs.[5]

5.3.4.1. Water Use Survey

A useful method to identify ways to reduce water consumption is a walk-through survey of the facility, in which a team walks the facility to observe and inventory inefficiencies related to water use. Such a walk-through survey enables the team to quickly locate "low-hanging fruit," items that save water at a small cost. This survey helps facilities to:

- Identify major sources of water loss, such as leaks, that can be repaired for significant savings
- Acquire a more complete understanding of the major end uses of water and flows of water in the facility that can be achieved solely by analyzing water utility bills
- Instill floor managers and workers with the importance of looking for and noticing leaks and inefficiencies, helping to create a culture of water efficiency

5.3.4.2. Water Balance

A water balance is a chart, table, or diagram of a facility's water use that shows the flows of water into and out of each process or operation in the facility. It typically includes every water-consuming component on-site and all flows out of the facility. A water balance can also include data on the cost of water.

When creating a water balance, it is important to make note of all water use in the facility and track water flows from the source through all operations at the facility to water losses and outputs including evaporative losses, input to products, and wastewater discharge. All uses of water in a facility should be included in a water balance. Common end uses of water include processing and auxiliary processing, heating and cooling, indoor domestic use, and landscape irrigation. The water balance should capture all of the flows into and out of each process at the facility and should verify

Table 5.1: Key Areas to Check During a Water Survey

Process and Equipment Use • Cleaning, Washing, Rinsing • Metal Finishing • Painting • Dyeing and Finishing • Photo Processing • Process Water Reuse • Product Fluming (Water Transport) • Pretreatment/Filtration systems • Pump and Conveyor Lubrication • Water Use in Products	Other Facility Support • Floor Washing • Air Emission Wet Scrubbers • Building Washing • QA/QC Testing • Laboratories • Landscaping and Irrigation • Dust and Particulate Emission Control • Decorative Fountains and Ponds • Vehicle Washing • Cooling Water for Air Compressors and Vacuum Pumps • Hazardous Waste Storage and Effluent
Cooling and Heating • Single-Pass Cooling • Cooling Towers • Boilers, Hot Water, Steam Systems • Air Washers • Boiler Scrubbers	
Sanitary and Domestic • Toilets • Faucets • Urinals • Showers • Faucets • Wash-up Basins	Kitchens • Food Preparation and Cleaning • Dishwashers • Ice Machines • Food Disposals

Source: U.S. Environmental Protection Agency.

that there is a balance between flows in and flows out. Benefits of compiling a water balance include:

- Ability to identify processes that present the best targets for reducing water use
- Establishment of baseline water use data, from which improvements in efficiency can be benchmarked
- Creation of a tool with which to compare different water efficiency improvement strategies

To complete a water balance, data on the volume of water flows for each process or operation in the facility should be collected, including:

- Water and sewer bills from the previous one to three years
- Numbers, sizes, and locations of water meters
- Sources of incoming water
- Diagrams of plumbing and water pipes, as well as irrigation schematics
- List of water-consuming processes
- Information about unsewered wastewater (that is, wastewater not discharged to municipal sewer systems, such as through separate permitted discharges)

5.3.4.3. Water Waste and Value Stream Mapping

Like water balances, value stream mapping can be a powerful tool for identifying water waste and savings opportunities. Value stream mapping is a method for creating a visual representation of the flows of information and materials across all the activities involved in producing a product for a customer. In a value stream mapping event, two maps are developed: a map of the current state showing key data about existing processes and a future state map showing changes that can be made to reduce waste. Value stream maps can be developed at the value stream level (e.g., for a product family) or at the process level after the team has selected a process to investigate more closely. By adding water data to value stream maps, value stream mapping can help with the following:

- Gaining a better understanding of where the greatest water waste occurs
- Identifying areas where improvements can be made to reduce excess water use and other wastes
- Developing an implementation plan for process improvements to reduce wastes in the value stream, prioritizing water efficiency projects alongside other lean and environmental projects
- Quantifying the expected savings from implementing those improvements
- Creating a culture of efficiency as team members, from supervisors to floor workers, become engaged in cutting waste

5.3.4.4. Root Cause Analysis

Root cause analysis is the process of identifying a problem and working through possible components to determine the most basic reason why the problem is occurring. Once the cause of the problem is identified, its recurrence can be prevented.

A useful root cause analysis tool is the "5 whys" technique. The approach uses a systematic questionnaire technique to search for root causes of a problem. The goal of the exercise is to pursue further until the root of the problem is identified. Completing a root cause analysis exercise should be a precursor to brainstorming potential solutions to water waste problems at the facility. By using such a tool, it can be ensured that the root cause of the problem is addressed first, instead of merely treating the symptoms.

An example of the 5 whys technique in relation to a particular industrial water usage scenario for cleaning parts is as follows:

1. Why are we using so much water?

The parts need to be cleaned before painting.

2. Why do the parts need to be cleaned?

The parts fail quality checks if they are not cleaned before being painted.

3. Why do painted parts fail quality checks?

The paint does not adhere when part surfaces are not prepared properly.

4. Why do the surfaces of the part need to be prepared?
The surfaces get contaminated by oils used in the previous process.

5. Why are oils used in the previous process?
The oils are used to prevent corrosion during storage.

5.3.5. Water Efficiency Improvement Strategies

Ways to reduce water use range from simple strategies, such as adjusting the flow of water or installing water-saving devices on equipment, to more involved options, such as reusing water or changing to a low-water or waterless process. There are five general types of water-saving strategies:[5]

- Adjust water flow
- Modify existing equipment or install water-saving devices
- Change to more water-efficient equipment
- Reuse or recycle water (treat if needed)
- Shift to a low-water or waterless process

When evaluating water reuse opportunities, it is important to consider both water quality and water quantity. Not all processes need the cleanest, highest-quality water. In many cases, it may be possible to reuse the wastewater from one process or operation as an input to another process or for another use at the facility (e.g., air handler condensate, reverse osmosis reject water), as long as the quality of water needed is matched for its intended use. Some testing and additional treatment of the process water may be needed to make sure it is acceptable for the next use.

5.3.5.1. Water-Efficient Cleaning Strategies

One key step that is relevant to reducing water use is managing water used in cleaning and rinsing. Many industrial and manufacturing businesses use large amounts of water to flush lines, rinse parts and tanks, and clean equipment, floors, and other areas. These cleaning and rinsing practices often are large sources of wasteful water use and, therefore, opportunities for water savings.

Use dry cleanup first:

- Use brooms, brushes, squeegees, and/or other tools to remove materials and debris in dry form before using water for secondary cleaning. (This saves water, reduces wastewater, and enables recovery of process materials.)

Eliminate unnecessary water use for floor washing:

- Sweep or use a water broom instead of hosing floors.
- Spot mop if necessary.

Use mistake-proof equipment:

- Use hoses that have automatic shut-off nozzles.
- Use efficient spray nozzles, high-pressure washers, and/or flow restrictors to clean efficiently while reducing water use. (High-pressure, low-volume sprays generally work better than low-pressure, high-volume sprays.)

Use efficient spray-washing and spray-rinsing techniques:

- Use water wisely, and turn off water when not in use.
- Do not use a hose as a broom; doing so wastes time, water, and energy.
- Optimize spray and rinsing techniques, and document the best practices in the standard work for the process.

5.3.5.2. Total Productive Maintenance

As a key strategy for finding and preventing water losses, as well as other equipment failures, total productive maintenance (TPM) is also relevant to water efficiency. TPM is a method that focuses on optimizing the effectiveness of manufacturing equipment. It focuses on team-based maintenance that involves employees at every level and function.

A key practice in TPM is to maintain equipment in a manner that enables workers to quickly identify and correct problems that may result in leaks or spills. Since many facility operations use water, not just production processes, it is important to extend the preventive maintenance practices of TPM to nonproduction areas (e.g., restrooms and irrigation) when looking for water issues.

Proactive Maintenance Tips to Reduce Water Use

- Adopt a user-friendly system for reporting water leaks, and fix leaks immediately.
- Inspect hot and cold water lines, steam lines and traps, water-using equipment, and plumbing fixtures routinely to identify potential problems and keep them operating properly.
- When performing maintenance on water-using equipment, replace worn parts and check to make sure that water-saving features (e.g., automatic shut-off valves) are operating properly.
- Shut off water supply to equipment in areas that are not in use.

5.4. Air Pollution Control

On a global scale, one of the most important impacts of industrial systems on the environment is the emission of gaseous, liquid, and particulate matter into the atmosphere. The easiest way to control air pollution is to eliminate it at the source; increasingly, the emphasis of plant-wide air pollution reduction programs has changed from emission reduction after the source, such as pollutant removal systems, to before-source reductions and reduction through environmental management. There are well-established air pollution control technologies and practices that greatly minimize the amount of air pollutants generated. These methods allow greater control and understanding of emission parameters and often lead to increased process efficiency.

Even with source reduction technologies, there will probably still be some air contaminants remaining that must be removed or destroyed. Industries do not have the option of sending these to a municipal treatment facility as they do with liquid or solid wastes. These emissions must be controlled at the industrial facility. Typically, air pollution control devices are divided into those that are used to control particulates and those that are intended for control of gaseous pollutants. Some processes may be capable of removing both types of contaminants.

5.4.1. Air Pollution Management Strategies

The key to reducing air emissions through management procedures is to use a well-developed pollution management plan with clearly defined goals. Every industrial plant and process is unique, and there is no single pollution management strategy that would be effective for every facility. Some important elements involved in designing and implementing a pollution management plan are as follows:

- Identification of sources of emission
- Review of applicable regulations
- Raw material review
- Waste characterization
- Quantification of emissions, including fugitive emissions
- Evaluation of accidental release potential
- Pollution control and waste minimization procedures
- Implementation and management of control procedures

The first and most important step in an air pollution management plan is the identification of the sources of pollution. To identify activities that contribute to air emissions, a thorough investigation of every process at a facility should be conducted. The investigation of each individual source should be conducted with personnel who have a good understanding of the process operations in question and the applicable regulations. With the sources identified, focused effort could be directed at managing the emissions without getting lost in other processes that do not contribute to emissions.

The investigation of each process should also include a review of current regulations applicable to the process. Regulatory requirements could come in the form of a specific permit for the process or a blanket regulation that affects the source. Operational and record-keeping procedures for a process and potential environmental impacts of the source are usually part of such a review. It is important to check with the permitting authority on any current updates or changes in the permit.

It is necessary to understand how the use of a raw material contributes to emissions. Apart from estimation of emissions, this allows easier evaluation of alternative raw materials that may provide reduction in emissions without adversely affecting cost or product quality. Raw material should also be assessed to determine how it affects other environmental aspects of the source. If not completely used by the source, methods of disposing of the excess material should be reviewed. Equally important is to review how the raw material contributes to the source's waste stream and if waste could be reduced through better material management and utilization. Process waste and any by-products also should be evaluated on how they are generated and disposed of. Volatile wastes could contribute to air emissions and need to be examined closely for storage and safety considerations. Reduction in total waste generation or alternative utilization through reuse or recycling could lead to reduced emissions.

Emissions should be reviewed to identify the materials that are emitted to the air and how each is controlled. Emissions are characterized by several techniques that include engineering equations, emission factors based on testing at similar sources, and emission tests performed on the specific source at the site. An emissions factor is a representative value that attempts to relate the quantity of a pollutant released to the atmosphere with an activity associated with the release of that pollutant. In most cases, these factors are simply averages of all available data of acceptable quality and are generally assumed to be representative of long-term averages for all facilities in the source category. The AP-42 Compilation of Air Pollutant Emission Factors, has been published since 1972 as the primary compilation of the EPA's emission factor information.[6] It contains emission factors and process information for more than 200 air pollution source categories. A source category is a specific industry sector or group of similar emitting sources. The emission factors have been developed and compiled from source test data, material balance studies, and engineering estimates.

Processes need to be investigated to determine the potential for unintended and irregular fugitive emissions or accidental releases of air pollutants to the environment. Governmental regulations require that the likelihood of such accidental releases and occurrences be analyzed and their quantitative impact on the environment and human health be determined so that mitigating steps can be planned and implemented. The EPA has developed a list of more than 100 substances that, in the event of an accidental release, could cause death, injury, or serious adverse effects to human health or the environment. If a facility has more than a threshold quantity of these substances in a process, then it must develop and implement a risk management program. That program must include a hazard assessment, a prevention program and an emergency response program. The EPA has set threshold quantities for each regulated substance based on its toxicity, reactivity, volatility, dispersibility, and flammability.

5.4.2. Air Pollution Emission Control

Stationary sources of air pollution emissions, such as power plants, steel mills, smelters, cement plants, refineries, and other industrial processes, release contaminants into the atmosphere as particulates, aerosols, vapors, or gases. These emissions are typically controlled to high efficiencies using a wide range of air pollution control devices. The selection of the appropriate control technology is determined by the pollutant collected, the stationary source conditions, and the control efficiency required. In some cases, pollutant emissions can be reduced significantly through process modifications and combustion controls. In most instances, however, some form of add-on pollution control equipment is installed in the ductwork (or flues) leading to the smoke stack to meet current allowable emission limits. Common methods for eliminating or reducing gaseous pollutants include the following:[7]

- Destroying pollutants by thermal or catalytic combustion, such as by use of a flare stack, a high temperature incinerator, or a catalytic combustion reactor
- Changing pollutants to less harmful forms through chemical reactions, such as converting nitrogen oxides (NO_x) to nitrogen and water through the addition of ammonia to the flue gas in front of a selective catalytic reactor
- Collecting pollutants using air pollution control systems before they reach the atmosphere

The most commonly used devices for controlling particulate emissions include:

- Electrostatic precipitators (wet and dry types)
- Fabric filters (also called baghouses)
- Wet scrubbers
- Cyclones

In many cases, more than one of these devices is used in series to obtain desired removal efficiencies for the contaminants of concern. For example, a cyclone may be used to remove large particles before a pollutant stream enters a wet scrubber.

Common control devices for gaseous and vapor pollutants include:

- Thermal oxidizers
- Catalytic reactors
- Carbon adsorbers
- Absorption towers
- Biofilters

References

1. "Pollution Prevention (P2)," U.S. Environmental Protection p2/Agency, www.epa.gov/index.htm, retrieved September 2014.

2. "Air Pollution and Clean Air Act," U.S. Environmental Protection Agency, www.epa.gov/air/caa/index.html, retrieved September 2014.

3. "Clean Water Act Summary," U.S. Environmental Protection Agency, www2.epa.gov/laws-regulations/summary-clean-water-act, retrieved September 2014.

4. "Estimated Use of Water in the United States in 2010," U.S. Geological Survey, http://pubs.usgs.gov/circ/1405/pdf/circ1405.pdf, retrieved September 2014.

5. "Lean & Water Toolkit," U.S. Environmental Protection Agency, www.epa.gov/lean/environment/toolkits/water/resources/lean-water-toolkit.pdf, retrieved September 2014.

6. "Air Pollution Emissions," U.S. Environmental Protection Agency, www.epa.gov/airquality/emission.html, retrieved September 2014.

7. "Air Pollution Emission Control Devices for Stationary Sources," Air & Waste Management Association, http://events.awma.org/files_original/ControlDevicesFactSheet07.pdf, retrieved September 2014.

Chapter 6

Waste Minimization and Material Recycling

Waste minimization signifies tactics that aim to eliminate waste through upstream interventions. On the production side, these strategies are emphasizing optimizing resource and energy use and lowering toxicity levels throughout manufacturing processes. These tactics refer to minimizing the waste and improving resource efficiency in or even before the manufacturing process or product design. They also focus on cleaner production, reuse of scrap material, improved quality control, and waste exchanges. On the consumption side, waste minimization strategies try to reinforce responsiveness, timely environmentally conscious consumption patterns, and consumer responsibility to reduce the entire levels of waste generation.

Minimizing waste often is accompanied by economic benefits such as applying inputs more efficiently to reduce purchases of raw materials. Manufacturers will notice a reduction in generated wastes as the volume of nonproduct outputs decreases. Supplementary cost savings can be realized through reduced expenditures for costly hazardous material management and disposal. Waste minimization plans can also be conducive to measurable successes in terms of market share, revenue growth, and cost savings. Robust material recycling programs frequently transfer themselves from a cost into a revenue stream when volumes warrant commodity rebates. Reductions in hazardous waste volumes can also have the effect of reducing the overall toxicity of the manufacturing process and final product, which can contribute to having fewer employees and consumers exposed to toxins and an entire improvement in workplace health. These factors typically result in increased employee satisfaction and retention, not to mention the reduction of the potential risks and liabilities related to the use, storage, and disposal of hazardous materials.

In this chapter, we mainly focus on lean manufacturing methodologies and solid waste minimization project plans for conducting waste minimization in a manufacturing system.

6.1. Lean Manufacturing Methodologies for Waste Minimization

The U.S. Environmental Protection Agency (EPA) defines lean manufacturing as a business model referring to a set of strategic methods that put emphasis on removing nonvalue-added activities (waste) while delivering quality products on time at minimum cost with better efficiency. In the United States, lean implementation is quickly growing throughout various manufacturing and service sectors to create competitive advantages. Whereas the concentration of lean manufacturing is on driving quick, continual improvement in cost, quality, service, and delivery, substantial environmental benefits usually ride the coattails or happen incidentally on account of these production-focused endeavors. Lean production methods often promote a culture of continuous improvement, employee empowerment, and waste minimization, very consistent with organizational and structural features advanced under environmental management systems and pollution prevention.

In 2000, the EPA, as an initial step in lean implementation, sponsored research on lean manufacturing that consisted of a series of case studies with the Boeing Company. The study concentrated on lean implementation at the Boeing Company, which resulted in substantial resource productivity enhancements with significant environmental improvement implications. In addition, the organizational culture in terms of continual improvement in recycling and waste removals was empowered by applying lean techniques at Boeing. This culture closely mirrored the organizational culture in successful environmental agencies that have been working productively to encourage development and promotion of environmental management systems, pollution prevention, waste minimization, design for the environment, and other voluntary initiatives. After the Boeing study, lots of other related research and projects have been conducted in this area, and many manufacturing systems have realized that implementing lean concepts and tools results in enhancements in environmental performance, even when lean activities were not initiated for environmental purposes. Lean methods emphasize continually improving the resource productivity and production efficiency, which often translates into less material, less capital, less energy, and less waste per unit of production. Lean manufacturing generates an operational and cultural environment that is highly contributing to waste minimization and pollution prevention. It can be leveraged to produce more environmental enhancement.

Lean manufacturing includes a primary paradigm shift from conventional "batch and queue" mass production to product-oriented "one-piece flow" pull production. While batch and queue contains the mass production of large lots of products in advance in accordance with potential or forecasted customer demands, a one-piece flow system reorganizes the production tasks in a way that processing stages of various kinds are performed instantly close to each other in a continuous flow. This shift includes highly controlled processes operated in a well-maintained, well-organized, clean environment that integrates codes of employee-involved, system-wide, and continual improvement. Whereas most of these techniques are interconnected and can happen simultaneously, most manufacturing systems begin by employing lean methods in a specific production area or at a "pilot" facility and then develop the use of the methods over time periods.[1]

Theoretically, the relation between lean production and environmental improvement is resilient. As we discussed earlier, the significant objective of lean systems is the systematic elimination of waste by emphasizing production costs, product quality and delivery, and worker involvement. For the entire system's levels, advanced manufacturing methods work to lower the resource intensity necessary to deliver a product or service to satisfy customer demands. This means that manufacturing systems implementing lean methods continually pursue reducing the materials, energy, water, space, and equipment required per unit of production. Although environmental endpoints, such as hazardous waste, air emissions, and wastewater discharges, are often not directly recognized in the types of manufacturing wastes targeted by lean initiatives, improvements in these areas are deeply embedded in the other kinds of manufacturing wastes. Table 6.1 lists seven common types of waste that lean manufacturing triggers to eliminate, accompanied by the environmental impacts that are related to each of them.

Simpson and Samson (2010) also researched environmental strategies for low waste operations. In their literature review, the authors highlighted how advanced manufacturing systems (i.e., use of practices such as lean, just-in-time, and total quality management) have a positive influence on environmental performance.[2]

Table 6.1. Seven Types of Manufacturing Waste Targeted by Lean Manufacturing and Their Related Environmental Impacts

Waste types	Examples	Environmental impacts
Defects	Production of off-specification products, components, or services that result in scrap, rework, replacement production, inspection, and/or defective materials	Raw materials consumed in making defective products; defective components, requiring recycling or disposal; more space required for rework and repair, increasing energy use for heating, cooling, and lighting
Waiting	Delays associated with stock-outs, lot processing delays, equipment downtime, and capacity bottlenecks	Potential material spoilage or component damage causing waste; wasted energy from heating, cooling, and lighting during production downtime
Overproduction	Manufacturing items for which there are no orders	More raw materials consumed in making the unneeded products; extra products may spoil or become obsolete requiring disposal
Movement	Unnecessary human motions or long-distance transportation	More energy use for transport; Emissions from transport; more space required for WIP movement, increasing lighting, heating, and cooling demand and energy consumption; more packaging required to protect components during movement
Inventory	Excess raw material, WIP, or finished goods	More packaging to store work-in-process; waste from deterioration or damage to stored WIP; more materials needed to replace damaged WIP; more energy used to heat, cool, and light inventory space
Unused creativity	Failure to tap employees for process improvement suggestions	Fewer suggestions of pollution prevention and waste minimization opportunities
Complexity	More parts, process steps, or time than necessary to meet customer needs	More parts and raw materials consumed per unit of production; unnecessary processing, increasing wastes, energy use, and emissions

6.1.1. A Comprehensive Framework for a Systematic Lean Implementation Methodology

We can summarize all the vital elements of lean performances into a single framework and advance a methodical lean implementation methodology in accordance with lean principle.[3] Here a basic and simple assessable continuous lean evaluation method has been presented. The proposed methodology is shown in Figure 6.1. The lean implementation methodology, which is shown in this figure, develops a typical road map regarding lean principle and lean tools together. The entire performance of the manufacturing processes can be assessed by taking into consideration the effectiveness

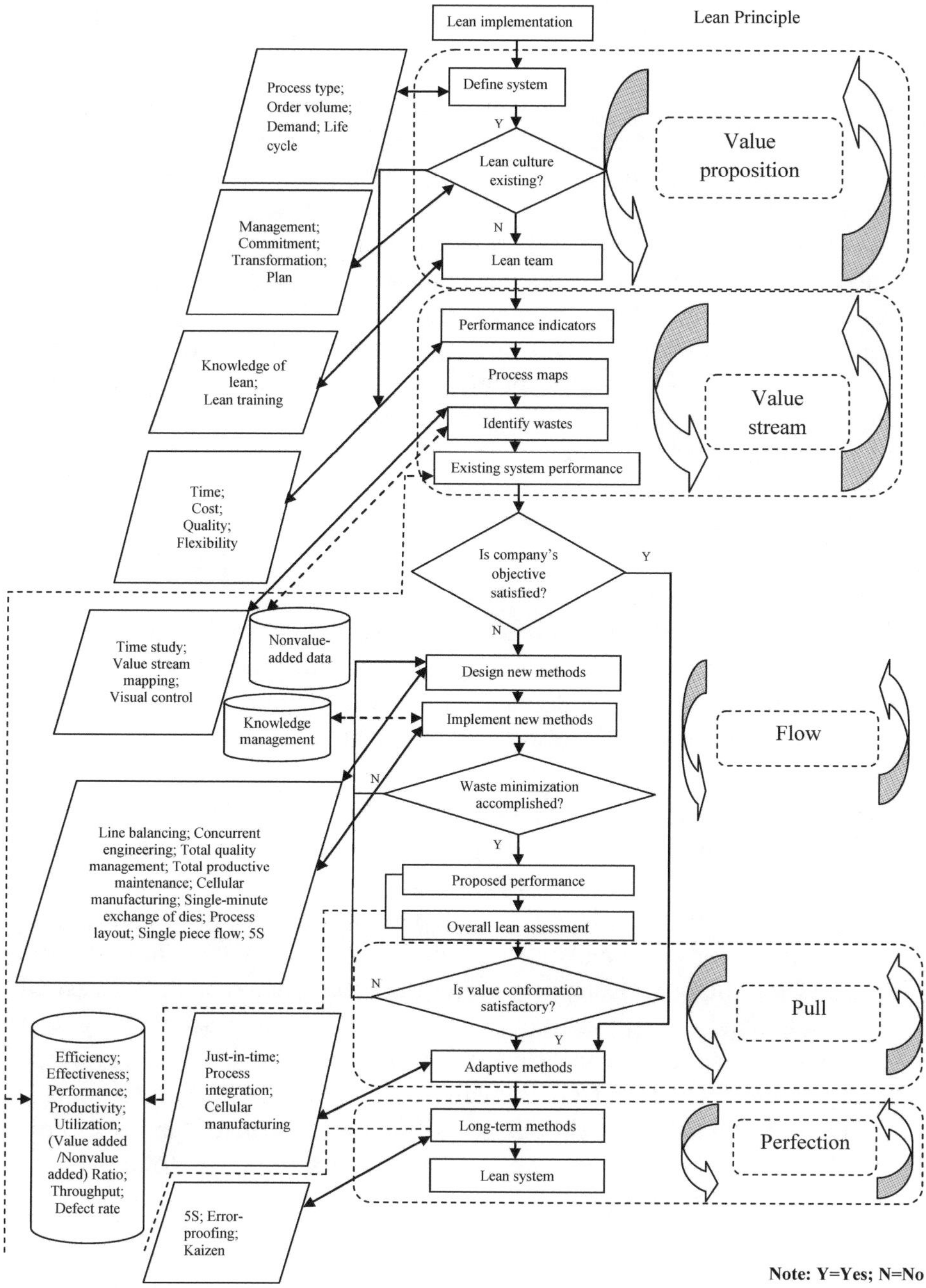

Figure 6.1. Lean Implementation Methodology

and efficiency of the production processes and waste reduction. The technique recommends that at an early stage of lean implementation, a lean team requires to be developed by experienced personnel from various sections of the manufacturing system. The team members should have adequate expertise in lean strategies and implementation methodology. Several companies and manufacturing systems may not have staffs with this expertise. A team member without knowledge and dexterity can unhelpfully affect the overall success of the project and can assess unrealistic performance for production processes. Recognition of central performance indicators is also important for leanness assessment. In some cases, it is not possible to recognize and analyze all the measures of performance indicators of a specific company. Ultimately, the implementation of this technique may be costly and time consuming, but new methods and techniques need to be utilized in consideration of a manufacturing system's budget and resources. Furthermore, a stage-by-stage description of the model, shown in Figure 6.1, is described in the steps below.

Step 1: Define the System. At the first step, manufacturers need to have access to a clear image and data regarding properties, parameters, and control variables of their system with consideration of production type, order volume, demand patterns, demand quantity, and amount of waste, since these indicators are highly associated with lean implementation.

Step 2: Lean Team. If there is not a lean culture in the manufacturing system, it can be initiated by forming a lean team based on a manufacturing system's policy, management commitment, and forthcoming plans. One of the initial actions about lean implementation cycles is lean training. The lean team should organize and conduct different training sessions related to the lean techniques, lean philosophy, implementation road map, essential knowledge and dexterities about lean. These training sessions provide a clear knowledge about lean's first principle, which is associated with value determination. The lean team typically contains various experts and management personnel from various sections, and their major target is to organize managers, staffs, and workers and control resources to implement lean in production processes.[3]

Step 3: Performance Indicators. In this phase, the lean team determines and measures the manufacturing performance variables based on the production and process features defined in the first step. The performance indicators can be evaluated in terms of cost, quality, lead time, processing time, operation time, and value-added time (value-added time is made up of processes that improve products). According to the literature, time-related measures are the most substantial measures in lean performance assessments and measurements.

Step 4: Process Maps. In this step, manufacturing processes are visualized and the value streams of that process are recognized by using visual controls, time study methods, and value stream mapping (VSM). As we described VSM in previous chapters (chapters 3 and 5), this tool is a lean method that utilizes a flow diagram documenting in high detail every stage of a process. Lots of lean specialists look at VSM as a structural tool to detect waste, reduce process cycle times, and accomplish process improvement. Some enterprises consider the value stream map as a characteristic of their lean efforts. VSM can be a vital method for documenting processes and removing wastes. Every process improvement initiative should start with an evident understanding of existing performance and an idea of the waste minimization the manufacturing system can attain. Value stream mapping is a lean management tool for evaluating the present status and planning a future status for a product or service during its life cycle.[4]

Step 5: Identify Wastes. This stage refers to the second principle of lean philosophy. Process maps represent all the existing wastes and nonvalue-added activities in a process that transforms raw materials into a finished good. Process maps also include flow of information and materials throughout the studied processes.

Step 6: Existing System Performance. A major part of the proposed methodology is to continuously assess performance before and after lean implementation. In this stage, the important objective is to measure the current state of the process in terms of productivity, efficiency, effectiveness, value-added time ratio, and defect rate using different lean assessment metrics. Moreover, it is necessary to assess the amount of waste in order to trace the results of improvement measures.

Step 7: Design New Methods (Lean Tools, Techniques, and Continuous Improvement). If the company's objectives regarding lean implementation are not satisfied, new design methods will be applied. Waste minimization is the core objective in this phase. Various lean strategies have been developed to reduce the nonvalue-added activities and enhance the leanness of manufacturing systems. However, the selection of lean strategies should be in such a way that implementing lean strategies should not increase other nonvalue-adding activities in the manufacturing process. Therefore, appropriate lean strategies must be selected to eliminate waste or improve the performance metrics in the manufacturing process. Moreover, it would be preferable to select the lean strategies that have the most overall impact on the identified wastes or performance metrics, according to the manufacturer's priority. As a result, applying the appropriate tool(s) at the right time within the budget for the right type of company is very important. Therefore, it is necessary to establish a proper relationship between the closely related lean strategies and manufacturing wastes.

The five key lean tools, especially suggested for process industries, are kanban, 5S (sort, set in order, shine, standardize, and sustain), visual control, poka-yoke (error-proofing), and single-minute exchange of dies (SMED). The most commonly used lean tools that are especially applicable in this step of designing new methods are 5S, SMED, total quality management, total productive maintenance (TPM), and cellular manufacturing. In cellular manufacturing, production workstations and equipment are organized in a sequence that supports a smooth flow of materials and components through the production process with minimal transport or delay. Implementation of this efficient method often represents the first major shift in production activity, and it is the key enabler of increased production velocity and flexibility, as well as the reduction of capital requirements.

Moreover, other useful techniques such as line balancing, concurrent engineering, and process layout can be utilized to make this step more efficient. Line balancing is a production strategy that involves setting an intended rate of production for required materials to be fabricated within a particular time frame. Effective line balancing requires assuring that every line segment's production quota can be fulfilled within the time frame using the available production capacity.

Concurrent engineering, also recognized as simultaneous engineering, is a technique of designing and developing products in which the different stages run simultaneously, rather than consecutively. It decreases product development time and also the time to market, leading to improved productivity and reduced costs. A process layout is a type of facility layout in which the floor plan is arranged with similar processes or machines located together. For instance, a machine shop with mills located together, lathes located together, and so on is arranged with a process layout. This differs from a product layout, in which equipment is arranged based on sequential steps involved in manufacturing a product, as on an assembly line.

Step 8: Implementing New Methods. Selection and implementation of new methods and tools strongly depend on the product volume and initial investment capability of the manufacturing system. It is challenging to identify the right lean tool based on the current production process of a company. Manufacturer's resource constraints should be considered to select and implement the right lean tools.

Step 9: Evaluation of Proposed Performance and Performing an Overall Lean Assessment. After selecting and implementing suitable lean tools, the performance and efficiency of the proposed process should be further evaluated. If value conformation is not satisfactory, adaptive methods will be applied by the manufacturing system.

Step 10: Adaptive Methods. Based on the fourth principle of lean, a production process should be organized like a pull system, where the wastes are removed from the process and the flow of the product is smooth with the least nonvalue-added tasks. A combination of various lean adaptive methods and useful techniques applied in industrial engineering, such as just-in-time, cellular manufacturing, and process layout, can be used in this step. Regarding any adjustment or improvement in the process, the manufacturer requires adopting an appropriate approach, hence, adaptive methods are one of the fundamental phases in the lean implementation project.

Step11: Long-Term Methods and Lean System. This step is associated with the establishment of long-term methods. In this phase, the culture for continuous improvement techniques is expanded, and one of the goals is that every employee related to the manufacturing processes will be interested in changing his or her mind-set in addition to working attitude toward lean systems or continuous improvement philosophy.

6.1.2. A Sample Research Study Related to the Application of Lean Manufacturing Methodologies in Waste Minimization

In this section, we focus on a research study conducted by Andrea Chiarini (2014). This study is related to empirical observation inside five European companies that manufacture motorcycle components and are also committed to lean and environmental management. The environmental impacts of the five companies' production processes were observed and measured before and after the implementation of three lean tools: 5S, cellular manufacturing, and TPM. Comparison of the before-and-after quantitative results revealed an interesting and novel outcome that contributes to the research on the effects of lean production on environmental impacts.[5]

- 5S can be useful for reducing oil leakage and improving waste management.
- Cellular manufacturing can lead to a decrease in electricity consumption.
- TPM can help to reduce the impacts of several machines, such as oil leakage and emissions of dusts and chemical fumes into the atmosphere.

The result of this empirical research also revealed other interesting positive effects concerning electricity consumption in general, as well as the standardization of activities and worker behaviors.

6.1.2.1. Implementation of 5S Methodology as One of the Lean Tools

The 5S methodology is usually structured by five precise steps. As we mentioned earlier in chapter 3, these steps are: sort, set in order, shine, standardize, and sustain. In each of the five companies, the observation took place in an assembly line where the 5S steps were applied in the same way; 5S had not previously been implemented in the five assembly lines.

Table 6.2 (on the following page) shows the activities put in place by the means of the 5S tool. In particular, the column on the left lists only those activities that positively affect the environmental impacts of the assembly lines. It is important to notice that the actions were the same for all the five companies and produced similar quantifiable benefits. Table 6.3 (see page TK) also refers to the achieved results and conducted measurements before and after implementation of the 5S methodology.[5]

Table 6.2. Environmental Benefits Observed by Applying Each of the Lean Tools

Lean tool	Activities	Environmental benefits observed
5S	Better identification and separation of the different kinds of industrial garbage (plastic, metal, cardboard, oiled rag, and solvent)	Reduction of mistakes during the separate collection of garbage in compliance with European Waste Catalogue coding; reduction of shop-floor space for manufacturing activities
	Adoption of a new standard drip tray for solvent and oil containers	Reduction of greased and solvent rags as a waste; reduction of leakage on the floor
Cellular manufacturing	Grouping the machine and equipment in a single cell	Reduction of material transportations toward the cell (this implies a saving of electricity for recharging the batteries of the electric truck)
TPM	Daily autonomous maintenance by the operators on the machine (lubrications, cleanliness of the filters, inspection of the instrumentations on the machine, etc.)	Relevant reduction of oil leakage on the floor, especially at the base of the machine
	Definition of a fixed frequency for maintaining or replacing critical parts of the machine	Reduction of dust and fume emissions, such as volatile organic compounds and ammonia

6.1.2.2. Implementation of Cellular Manufacturing Methodology

Cellular manufacturing is a lean tool for grouping machines, workplaces, and staff dedicated to similar products in a single cell, rather than having them in different places on the plant layout. By means of a particular *U*-shape, the workplaces and machines in the cell are placed very close to one another to save space and time. The handling and transportation of materials can, in this way, be easily reduced. All five companies implemented similar cells in many points of the plant layout, achieving in this case a reduction of energy consumption of the electric trucks. Indeed, having grouped different machines and workplaces, the transportation path among

Table 6.3. Measurement of the Environmental Benefits Before and After Implementation of Lean Tools

Lean tool	Impact	Before implementation of the lean tool							After implementation of the lean tool						
		A1	A2	A3	A4	A5	Mean (μ)	Standard deviation (σ)	A1	A2	A3	A4	A5	Mean (μ)	Standard deviation (σ)
5S	Greased rags (kg/month)	1.1	2	1.8	2.2	1.7	1.76	0.416	0.3	0.4	0.6	0.3	0.3	0.38	0.130
	Rags for solvent (kg/month)	0.8	0.7	1.9	2	1.1	1.43	0.629	0.1	0.2	0.2	0.2	0.2	0.18	0.045
	Oil leakage on the floor (L/month)	3.8	4	3.1	3.9	1.1	3.18	1.215	0.4	0.3	0.1	0	0.1	0.18	0.164
Cellular manufacturing	Electricity consumption of the truck (KWh/month)	121	116	117	128	110	118.4	6.656	38	41	36	39	40	38.8	1.923
TPM	Oil leakage on the floor (L/month)	5.1	3.8	2.6	4.8	4.2	4.1	0.980	0.3	0.2	0.4	0.4	0.4	0.34	0.090
	Dispersed dusts in the layout (mg/m^3)	9.1	9	6.4	3.1	2.3	5.98	3.196	2.3	2.4	2.2	1.9	2	2.16	0.207
	Volatile organic compounds (mg/m^3)	11.4	14.2	14	11.6	17.3	13.7	2.398	9	8.9	9	8.8	9.1	8.96	0.114
	Ammonia emissions (mg/m^3)	6.7	6.4	6.3	12.4	4.5	7.26	3.001	4.1	4.1	3.9	5.2	4.3	4.32	0.512

Note: A1–A5 refers to the indices of five selected sample companies.

the stations was drastically reduced. The results implied a less frequent recharge of the batteries of the electric trucks and ultimately a reduction in electricity consumption. Based on the information demonstrated by Table 6.2 and Table 6.3 earlier, a reduction of 68% in terms of kilowatt hours, due to battery recharge in a month for the first company, was achieved. The implementation of cellular manufacturing in the five companies also demonstrated a reduction of the environmental impacts, in particular, the electricity consumption.

6.1.2.3. Implementation of TPM Methodology

TPM is a complete management system made up of different tools and principles invented in Japan in the 1960s for reducing machine stoppages in particular owing to failures. TPM is managed at two important levels. The first and basic level is called autonomous maintenance. Operators every day have to clean the machine and perform basic maintenance, such as topping up oil levels, cleaning filters, and checking pressure, temperature, and so on of the machine. At a more engineering level, supervisors collect data related to the failure frequencies for setting a preventive maintenance program. All five companies had never before implemented TPM. A press for plastic products was the first machine in which TPM was implemented. After six months of implementation, the companies observed and measured several positive effects. The actions taken for implementing TPM were carried out at the same time. For instance, while the operators started autonomous daily maintenance of the machine, the engineers collected data and decided the frequencies of maintenance and/or replacement of the critical parts of the machine. In this way, all these actions led to important benefits that help control negative impacts on the environment.

The benefits can be classified into two specific categories. The first is related to the reduction of leakage from the machine to the floor. Effective daily maintenance by the operators and scheduled maintenance to the parts of the machine in these companies dramatically reduced the leakage level to almost zero. In addition, all the companies observed a reduction of the atmospheric emissions in terms of dusts and fumes. The press for plastic products usually heats up a plastic compound, melting it in the mold. This produces emissions through the chimneys of dusts, volatile organic compounds, and ammonia. All the companies were already below the targets required by laws. The emissions decreased, and this was particularly due to the more effective maintenance of the filters, piping, and chimney of the machines. Some managers of the five companies also argued that TPM could affect equipment energy consumption, especially in the cells. Indeed, they noticed that TPM helps to turn off the energy users in a cell and in the equipment in general. This will lower nonvalue-added standby energy use, although without affecting standby time. This is an interesting improvement introduced by TPM that the managers are quite sure has to be taken into consideration (Table 6.2 and Table 6.3).[5]

6.2. Solid Waste Management

Recycling and solid waste minimization in manufacturing systems can be accompanied by various environmental, economic, and social benefits. The greatest environmental benefits of recycling and solid waste minimization are related to conservation of energy and natural resources and the prevention of pollution, owing to the fact that a recycled material rather than a raw material will be used in order to make a new product. As recycled materials have been refined and processed once, manufacturing the second time around is much cleaner and will need less energy than the first time. The key benefits of solid waste minimization can be categorized as conservation of natural resources, healthier environment via landfill emission reduction (carbon dioxide, methane, and leachate), global warming reduction, and conservation of habitats.

Moreover, recycling and solid waste minimization benefit the manufacturing system economically through three different channels: cost savings in material purchases and energy consumption because of reuse and reduction, cost avoidance of solid waste hauling and disposal, and revenue generated from the sale of recyclable material. For instance, the Business Waste Reduction Assistance Program at the University of Toledo (Ohio) recently identified over $3.1 million in annual savings for Northwest Ohio businesses in the 70 waste assessments the program has completed. For example, at a plastics manufacturer with 100 employees, approximately $16,000 in annual cost benefits was identified via increased plastic, paper, and cardboard recycling. The waste stream amounts and revenues also cost-justified the purchase of a baling system[6, 7, 8, 9].

Solid waste minimization also offers personal and social benefits—although many of these benefits are somewhat intangible and somehow difficult to measure—such as sustainable environment for future generations, cleaner facilities, buy-in at work programs, healthier environment for working and living, and money generation to assist local programs.

6.2.1. A Sample Plan for a Solid Waste Minimization Project

Here a sample plan for a solid waste minimization project using the six sigma deployment plan based on the DMAIC (Define, Measure, Analyze, Improve, and Control) cycle is presented. Six sigma is an inclusive and flexible method for achieving, sustaining, and maximizing enterprises' success. It is distinctively driven by a close understanding of customer requests; well-organized use of facts, data, and statistical analysis; and thorough attention to managing, improving, and reinventing system processes. Also, six sigma attempts to enhance quality, productivity, and bottom-line financial performances in a system. The six sigma methodology provides a manufacturing system with a base for institutionalizing improvement initiatives such as an effective solid waste minimization program to achieve defined goals and successes.

It is also worthwhile to mention that six sigma counts greatly on data, facts, and the use of statistical tools to evaluate whether an improvement has been made. Six sigma methodologies are categorized into six essentials, mentioned below:

- Defining products or services
- Knowing the stakeholders and customers and their critical needs
- Identifying processes, methods, and systems to meet stakeholders' critical needs
- Establishing a process for doing work activities consistently
- Error-proofing the processes and eliminating wastes
- Measuring and analyzing the system performance

To achieve these fundamental objectives, six sigma utilizes the five-step DMAIC methodology. Moreover, six sigma has three overall targeted solutions: process improvement, process design/redesign, and process management.

6.2.1.1. Background of the Company and the Related Assumptions

The company, selected for conducting a waste minimization project, is a metal fabrication and precision tooling facility that generated $8,200,000 in annual sales; after subtracting all of the costs (production costs, costs of purchasing raw materials, overhead costs, transportation costs, disposal costs, etc.), equal to $7,400,000, $800,000 is left as profit. The company employs 620 people with full engineering and maintenance teams. It currently does not have any recycling programs and spends $450,000 for waste removal. This information shows that ($\frac{\$450{,}000}{\$7{,}400{,}000} = 0.06$) 6% of the total cost is related to the waste removal, and it is a significant amount. If this company would be able to reduce this amount, and also generated revenue by purchasing the recyclable materials, the outcome would be more efficient in terms of economic viewpoint. Such a strategy would be accompanied by environmentally friendly action. Based on the gathered data, we have the following data related to the waste materials: Tables 6.4–6.8).

Table 6.4. Disposal Materials and Their Related Amounts for the Sample Company

Disposal materials	Amount (metric ton per year)
Office paper	3,000
Food waste	2,500
Cardboard	2,000
Wood	1,500
Steel	1,000
Total	10,000

Table 6.5. Waste Materials and Selling Price of Recyclable Materials for the Sample Company

Waste materials	Selling price of recyclable materials (dollars per metric ton)
Office paper	75
Food waste	No customer to sell them
Cardboard	100
Wood	No customer to sell them
Steel	600

Table 6.6. Waste Materials and Transportation Costs for Recyclable Materials for the Sample Company

Waste materials	Transportation costs of recyclable materials (dollars per metric ton)
Office paper	35
Food waste	No customer to sell them
Cardboard	35
Wood	No customer to sell them
Steel	200

Figure 6.2 shows the related Pareto diagram. While guiding the solid waste management project, after the baseline data have been analyzed, the assessment team can start investigating individual elements in the waste stream that should be taken into account for reduction, reuse, or recycling. A helpful technique to implement this task is to perform a Pareto analysis. A Pareto analysis is a statistical method of decision making that is applied for the selection of a restricted number of activities that generate the substantial overall effect. The Pareto principle is also famous as the 80/20 rule, in that 80% of the effects are caused by 20% of the causes. The concept is to identify the 20% significant waste components that generate 80% of the total waste and then target the 20% significant causes for waste minimization.

Table 6.7. Collection Costs of Recyclable Materials for the Sample Company

Waste materials	Collection costs of recyclable materials (by employing some staff and required equipment) (dollars per metric ton)
Office paper	30
Food waste	No customer to sell them
Cardboard	30
Wood	No customer to sell them
Steel	170

Table 6.8. Computation of Annual Disposal Costs for the Sample Company

Waste materials	Portion of each kind of waste material (in disposal amount per year)	Annual disposal cost
Office paper	$^{3,000}/_{1,0000} = 30\%$	$450,000 × 0.30 = $135,000
Food waste	$^{2,500}/_{10,000} = 25\%$	$450,000 × 0.25 = $112,500
Cardboard	$^{2,000}/_{10,000} = 20\%$	$450,000 × 0.20 = $90,000
Wood	$^{1,500}/_{10,000} = 15\%$	$450,000 × 0.15 = $67,500
Steel	$^{1,000}/_{10,000} = 10\%$	$450,000 × 0.10 = $45,000

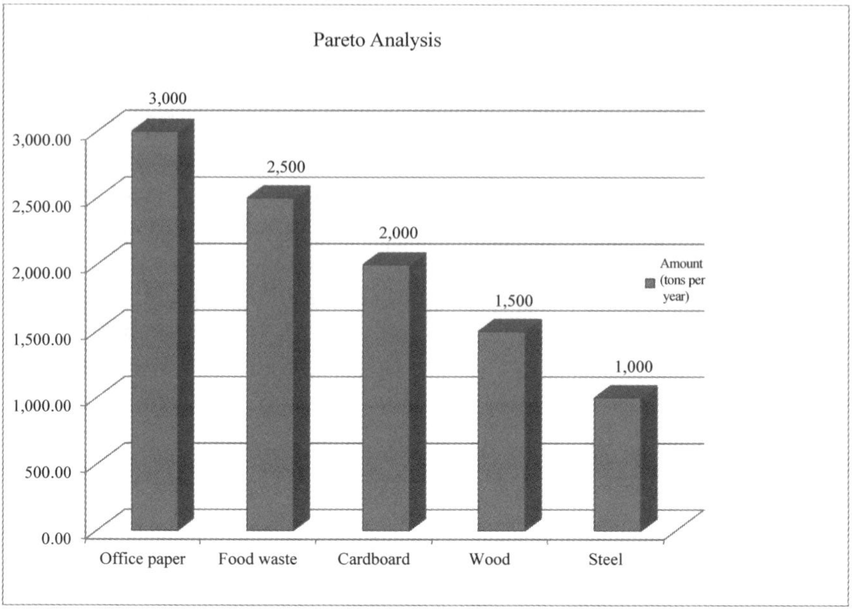

Figure 6.2. The Pareto Diagram of Recyclable Materials

Reviewing the Existing Organizational Records for Solid Waste and Recycling: Reviewing the existing manufacturing system's records for solid waste and recycling typically gives significant insight into the amounts, kinds, and forms of waste generation. Some suitable data and information can be collected to assist in focusing on the team's efforts and eliminating the need for the collection of existing raw data. A great deal of time, money, and effort will be saved by gathering high-quality records versus raw data collection via a trash sort or site walk-through. Various types of records that are useful to be collected consist of:[10]

- Purchasing, inventory, maintenance, and operating logs
- Supply, equipment, and raw material invoices
- Equipment service contracts
- Repair invoices
- Waste hauling and disposal records and contracts (including annual amounts and fees collected)
- Contracts with recycling facilities and records of earning revenues from recycling

- Major equipment list
- Production schedule (for one year)
- Company brochure or product information
- Material safety data sheets
- Facility layout
- Process flow diagrams

Moreover, the purpose of reviewing these records is to determine the total amount of waste generated per year, the total amount of material recycled per year, and the total waste removal and recycling cost structure. A review of purchasing records can be beneficial to backpedal into an estimate of waste components generated. For instance, shipments often arrive in cardboard containers. Finding the number of containers received per year and appraising the weight per carton can assist the team in assessing the total weight of cardboard boxes disposed of annually. In addition to what has already been written, the janitorial, maintenance, purchasing, and accounting staffs will be most helpful when gathering these records. Customer service at the waste hauling and recycling companies may also record more thorough records than the information that exists on the monthly bills.

Furthermore, it is worthwhile to mention that the team should be aware of some constraints that exist from the existing records, for example:

- The records may not provide sufficient data if the waste hauling company does not collect this information.
- The records do not typically provide information about particular waste components (a trash sort is generally needed for these data).
- The records can be hard to use if more than one enterprise/company shares a dumpster.

6.2.1.2. Initial Sorting for Solid Wastes and Conducting a Composition Study

Solid waste sorts offer detailed data regarding the composition of a manufacturing system's waste stream. Using the data collected from sorting the solid waste, the manufacturing system's waste stream can be typified into the different materials that make up the whole stream, comprising the annual generated amounts and the percentage that each component contributes to the entire waste stream. These data are very useful when assessing cost-effective methods or process changes to divert these components from landfills. Sorting the waste itself is tenderly referred to as "dumpster diving," since the team will tangibly collect, sort, and weigh a representative sample of the manufacturing system's waste. In this stage, the data collection form is one the key documents of sorting the waste. The data collected by the form will be applied to extrapolate the annual generation for the facility, so care should be taken when collecting the data to ensure accuracy. At least, the required information on the data collection form is as follows:[10]

- The date that the data were collected
- The team members collecting the data (this is very useful if follow-up is needed)
- Work unit and location of the waste receptacle
- Source of the waste (previous operation or supplier)
- Disposal method (baler, compactor, recycler, uncompacted dumpster)
- The size of the container in $feet^3$ or $meters^3$ (can be derived from length, width, and height measurements)
- The container type (desk side, recycling bin, dumpster)
- Percentage full
- Times emptied (per day, week, or month)
- Container contents and percentage of each component
- Condition of material (loose, compacted, baled)
- Notes and comments that may be helpful when evaluating the data (containing names and contact information for work unit members with particular knowledge of the waste or generation levels)

Moreover, we should consider that before scheduling the actual sorting of the waste, several preparation working activities are needed, consisting of various team meetings to discuss assorted items to ensure the smooth implementation of the project. Particularly, the team should focus on the following items to finalize earlier than the sort:

- Holding a team meeting
- Creating a waste assessment plan and assigning roles and facility regions of responsibility
- Conducting waste sort training sessions
- Reviewing the layout of the facility
- Gathering the needed tools and equipment
- Assigning data collection roles to temporary team members
- Preparing the data collection forms
- Holding a custodial staff meeting
- Disseminating employee communications concerning the sort and what to expect

Sorting the waste itself is straightforward; every individual waste and recycling receptacle in the facility is evaluated and a data collection form is completed. During the walk-through, the assessment team members will collect raw data on what they watched as well as record the location and the major use of each dumpster. The team inspects the company's main production, the waste flow in the facility, and the number of waste streams in the facility, and it depicts a process and material flow diagram. Ideally, if a layout of the facility is accessible, it may be applied to segment the facility to collect data for the team members.

6.2.1.3. Scope/Goal of the Project (What Areas Will Be Targeted?)

The major objectives of this waste minimization project can be categorized as follows:

1. Identifying major solid waste streams and suggesting potential areas for reducing, reusing, and recycling materials in the waste stream (determining recycling opportunities and cost savings for the metal fabrication company)
2. Reducing the cost and the amount of material related to solid waste disposal and making revenue by the proposed recycling plan and increasing the efficiency

3. Providing this metal company with some alternatives (reuse, recycling, and composting alternatives) and recommendations in order to minimize waste and waste management effectively

The areas that will be considered as our main target are related to recyclable materials that can be sold in special markets. In this process, we try to identify recyclable materials and evaluate the potential profit that can be achieved by selling them and avoiding disposal as much as possible.

Based on a conducted initial sorting of solid wastes, a rapid inspection process can assist in speedily distinguishing the worthwhile options for full assessment and possible compositions in the waste minimization program. In addition, nonefficient options can be eliminated, saving the team time and funds in the assessment process. An efficient inspection process should be in association with the original defined objectives for the project and at least should provide the following information:[10]

- Expected solid waste reduction (annually in tons)
- Expected start-up costs
- Impact on waste removal costs (annually in dollars)
- Impact on purchasing costs (annually in dollars)
- Impact on employee morale
- Ease of implementation

The team should consider that the objective of the inspecting process is to rapidly identify options worthy of further assessment. A weighted scoring system can be applied to regularly rank each alternative in an objective manner. A quality deployment function, such as the House of Quality, is an excellent technique to implement this assessment. The House of Quality generally functions as a living document and a source of ready reference for related products and future upgrades. Its purpose is to serve as a vehicle for dialogue to strengthen vertical and horizontal communications. Through customer needs and competitive analysis, the House of Quality helps to identify the critical technical components that involve changes. Subjects are addressed that may never have surfaced before. These critical matters are then driven through the other matrices to identify the critical parts, manufacturing operations, and quality control measures needed to produce a product that fulfills both customer needs and producer needs within a shorter development cycle time. The net effect of all of this is that the items that drive the manufacturing system's actions are spurred by customer requirements. There is a rising focus on the customer and an improved awareness of the wants of the customer. Owing to this emphasis, the process leads to enhanced customer understanding and the ultimate outcome for having more satisfied customers.

In addition, before proceeding with the inspection process, the team should determine the assessment criteria and weighting system. A scale of 1 to 10 for weighting each criterion is suggested. These weightings should be defined by group decision making of the team, project manager, facility manager, and sustainability manager. The assessment criteria should be directly associated with the objectives defined for the project (e.g., decrease in waste amounts, reduction in waste toxicity, reduction in waste disposal costs, reduction in purchasing costs, revenue generation potential, low start-up costs, productivity enhancement, quality improvement, ease of implementation, impact on employee morale, impact on company image, and impact on safety).

6.2.1.4. Methodology: Proposed Six Sigma Deployment Plan Based on the DMAIC Cycle (How Will We Start?)

As the needs of the company for minimizing solid waste and doing a waste minimization project have been identified and chief managers support and assign resources to the effort, evaluations can be performed based on applying six sigma methods and the DMAIC cycle.

In general, the waste minimization deployment plan includes 10 divisions:

1. Cover page with official approvals
2. Overview
3. Assumptions, dependencies, constraints, and limitations
4. Operational readiness
5. Timeline and schedule for implementation
6. Training and documentation
7. Notification of deployment

8. Operations and maintenance plan
9. Contingency plan
10. Appendices

Moreover, some criteria should be evaluated when we are exploring the technical feasibility of a waste minimization option. Technical and operational feasibility studies are associated with this matter, whether the suitable resources exist or are practically achievable to implement a particular option or not. This encompasses the square footage of the building, existing and accessible utilities, existing processes and material-handling equipment, the quality requirements, and the proficiency level of employees. Alongside this process, product specifications and facility constraints should be considered. Usual technical evaluation criteria consist of:

- Available space in the facility
- Safety
- Compatibility with current work processes and material handling
- Impact on product quality
- Required technologies and utilities (power, compressed air, and data links)
- Knowledge and skills necessitated for operating and maintaining the alternative
- Additional staff and labor requirements
- Impact on product marketing
- Completion and execution time

Once the technical feasibility is assessed, with the help of facility engineers or consultants, inputs can be initially analyzed. It is wise to discuss the technical aspects with staffs and workers who are directly impacted by the change, for instance, staffs and workers who work in production and maintenance sections. If an alternative requires an adjustment in raw materials, the effect on the quality of the final product should be assessed. If an alternative does not satisfy the technical requirements defined in the system, it should be eliminated from consideration. From a technical viewpoint, the three regions that need additional assessment are:

- Modifications in equipment, machines, purchases
- Changes in processes
- Changes in materials, raw materials, and substances

To ensure the defined goals are met, using a systematic framework that applies the five-step six sigma/DMAIC process can be really efficient. This framework is illustrated by Figure 6.3.

In the first step, we will conduct a meeting and attempt to launch the solid waste minimization team and charter and assign the task leaders. In this meeting, we will try to visit seven core members representing different regions in the facility, consisting of company manager, production engineer, maintenance technician, production supervisor, line employee, janitorial worker, and accountant. Then we consider an introductory training meeting with our team in order to exchange information/work schedules, establish

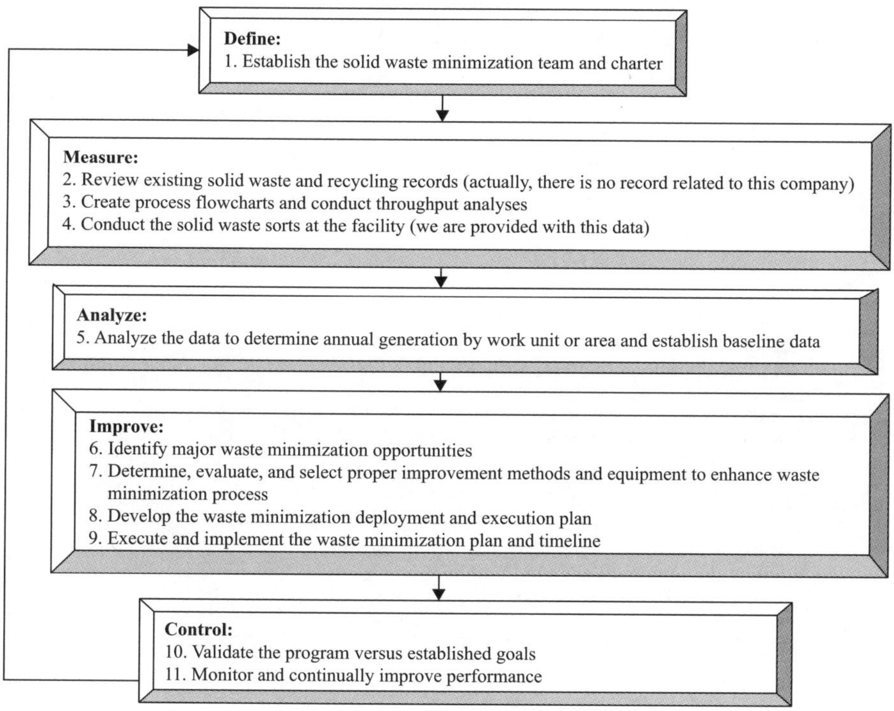

Figure 6.3. Six Sigma/DMAIC Process Flowchart

comprehensible SMART (specific, measurable, achievable, realistic, and timely) goals, outline initial team member roles, confirm the budget and the timeline, and create the team charter and work break-down structure.

6.2.1.5. Project Scheduling (How Long Will the Project Last?)

The timeline and Gantt chart for this project appear in Table 6.9 and Figure 6.4, respectively.

Table 6.9. Timeline for the Solid Waste Minimization Project

Project start date: 6/20/2017 (Monday)				
Activity #	***Description***	***Start***	***End***	***Duration (working days)***
1	Establish the solid waste minimization team and charter	6/20/17	7/04/17	11
2	Review existing solid waste disposal records (this company doesn't have such recycling records, but we have the data regarding the disposals)	7/04/17	7/11/17	5
3	Create process flowcharts and conduct throughput analyses	7/11/17	7/25/17	10
4	Conduct the solid waste sorts at the facility	7/25/17	8/08/17	10
5	Analyze the data to determine annual generation by work unit or area (we have some data related to that)	8/08/17	8/22/17	10
6	Establish baseline data	8/22/17	8/29/17	5
7	Identify waste minimization opportunities	8/29/17	10/03/17	25
8	Develop process and method improvement alternatives to minimize solid waste	10/03/17	11/07/17	25
9	Determine vendors and service providers to assist in waste minimization	11/07/17	11/21/17	10
10	Compare and decide among alternatives	11/21/17	11/28/17	5
11	Develop the waste minimization deployment and execution plan	11/28/17	12/12/17	10
12	Execute and implement the waste minimization plan and timeline	12/12/17	1/09/18	20
13	Validate the waste minimization process versus goals	1/09/18	1/16/18	5
14	Monitor and continually improve performance	1/16/18	1/23/18	5
Total time of doing this project (working days) = 156				

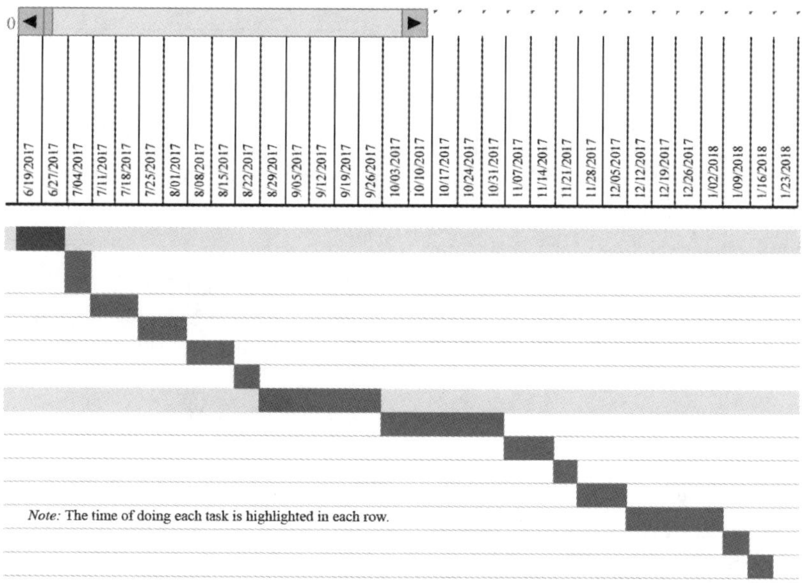

Figure 6.4. The Gantt Chart of Solid Waste Minimization Project Based on Defined Activities

6.2.1.6. Expected Benefits (What Are the Waste Reduction and Cost Benefits?)

In this section, type of waste reduction, annual benefits, annual costs, and annual total profits are calculated and the results presented in Tables 6.10 and 6.11.

Table 6.10. Computation of Annual Revenue, Annual Cost of Collection, and Annual Cost of Transportation Regarding Recyclable Materials

Waste materials	Annual revenue by selling them in dollar	Annual cost of collection (by employing some staff and required equipment) in dollars	Annual cost of transportation in dollars
Office paper	3000 × 75 = 225,000	3000 × 35 = 105,000	3000 × 30 = 90,000
Cardboard	2000 × 100 = 200,000	2000 × 35 = 70,000	2000 × 30 = 60,000
Steel	1000 × 600 = 600,000	1000 × 200 = 200,000	1000 × 170 = 170,000

Table 6.11. Computation of Annual Total Profit Regarding Recyclable Materials

Waste materials	Annual achieved benefit by selling recyclable materials in dollars	Annual cost savings (as they are not disposed of) in dollars	Annual total profit achieved by selling and cost savings (as they are not disposed of anymore) in dollars
Office paper	225,000 – (105,000 + 90,000) = 30,000	135,000	30,000 + 135,000 = 165,000
Cardboard	200,000 – (70,000 + 60,000) = 70,000	90,000	70,000 + 90,000 = 160,000
Steel	600,000 – (200,000 + 170,000) = 230,000	45,000	230,000 + 45,000 = 275,000

Annual total profit achieved related to selling recyclable materials and cost savings of disposal (as they are not disposed of anymore) is equal to:

(165,000 + 160,000 + 275,000) = $600,000

6.2.1.7. *Application of Lean Six Sigma Method, Statistical Process Control, Hypothesis Tests, and Hypothesis Flowchart in Waste Minimization and Controlling the Process*

Here we focus on the application of various statistical methods in waste minimization and process control.

Lean Six Sigma Method: Based on the conducted analysis, the company can expect a significant amount of benefit by selling some parts of solid wastes and cost savings that are originated by not disposing recyclable materials. Also, regarding the food waste, the company solid waste minimization manager or an audit team can conduct a feasibility study regarding the compost. Moreover, by applying the lean six sigma/DMAIC method, they can focus on a target for the amount of waste generated in the company and define lower and upper specification limits and try to control variations between them (Figure 6.5). In this way, the outcomes will be more efficient in terms of both economic and environmental perspectives.

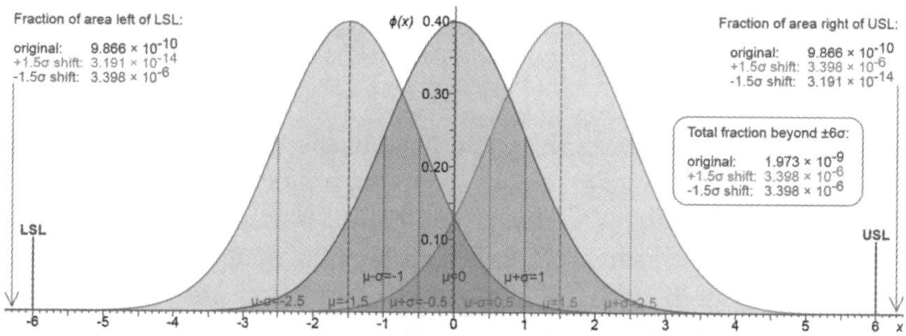

(Note: LSL and USL refer to lower specification limit and upper specification limit successively.)

Figure 6.5. An Example for Setting Targets and Applying Six Sigma Method to Control Waste Generation and Production Process in a Manufacturing System

Statistical Process Control: Statistical process control (SPC) procedures assist to monitor various process behaviors. One of the most successful SPC tools is the control chart, originally advanced by Walter Shewhart in the early 1920s. A control chart provides a record of data and informs us when an unusual and out-of control occurrence (e.g., a very high or low observation in comparison with usual process performance) happens. Control charts are applied to recognize variation, whether it is random or assignable. Control charts record an activity over a given period and are plotted against the specification limits to examine if the process is in control. Control charts try to identify two kinds of process deviations:

- Ordinary cause variation, which is inherent in the process and will always be present
- Distinct cause variation, which originates from external sources and specifies that the process is out of statistical control

Control charts usually have three design parameters: the sample size (n), the sampling interval (t), and the width coefficients of warning and control limits (w, k). These are the design parameters of the control chart. Various tests can assist to decide when an out-of-control occurrence has happened—though as more tests are used, the probability of a false warning also increases. Several SPC methods have been revived by American firms in recent years, especially as an element of quality improvement initiatives such as six sigma. More recently, others have supported assimilating SPC with engineering process control tools, which regularly change process inputs to enhance performance.

Hypothesis Tests: Performing a hypothesis test is also one of the key components in analyzing data for a lean six sigma project. A statistical hypothesis is a claim or conjecture concerning one or more samples of a related population. Testing of a hypothesis is an inferential method applied to make a statement about an activity or process on the basis of its output. Once testing the hypothesis is done, one of the following states can occur. The first one refers to rejection of the hypothesis, and the second one denotes that the applied test can be failed in rejection of that hypothesis. *Fail to reject* denotes that the data do not give adequate proof; in contrast, *rejection* means the sample testimony refutes the hypothesis. Rejecting the hypothesis signifies that when the hypothesis is true, there is a small probability of that sample occurring once more. There are usually two types of hypotheses: the null and the alternative. The test is designed against the null; if the test is rejected, then the alternative is confirmed. When doing this method, it is important to carefully choose the appropriate test and to interpret the results accurately. Hypothesis testing is an intimidating process, especially for those unfamiliar with statistics. This can be made easier by breaking down the process. One of the first steps in choosing the right hypothesis test, and for the lean six sigma methodology in general, is ensuring whether or not the data are normal. Those new to hypothesis testing tend to assume data normality, as that simply makes data analysis easier. However, there is danger in these assumptions, as certain projects may involve typical data (such as process cycle-time reduction).

Moreover, it is worthwhile to mention that normal data have normal variation and generally take the shape of a bell curve. This curve represents the data's central tendencies. The properties of this type are such that those working with it may use a probability plot for the actual verification that it is distributed in a normal way. The probability plot is a graphical technique for assessing whether or not a data set follows a given distribution, such as the normal distribution or Weibull distribution. The data are plotted against a theoretical distribution in such a way that the points should form roughly a straight line. Departures from this straight line signify departures from the specified distribution.

The correlation coefficient associated with the linear fit to the data in the probability plot is a measure of the goodness of the fit. Estimates of the location and scale parameters of the distribution are given by the intercept and slope. Probability plots can be generated for several competing distributions to see which provides the best fit, and the probability plot generating the highest correlation coefficient is the best choice, since it generates the straightest probability plot.

The probability plot serves as the first step in deciding which hypothesis testing is indicated. Once this information regarding the normality of the data is determined, it is necessary to determine the type of test to use.

Hypothesis Flowchart: The hypothesis flowchart is also used to give direction regarding the type of test. This chart encourages users to input the normal data from each "line" or "process" examined. If the comparison is being done between two or more groups, the variances of each line are examined. One of the tests that help determine the variance is the test for equal variance (F-test). Lines with equal variances should then use the analysis of variance (ANOVA) for hypothesis testing.

On the other hand, if the data are not normal, users can try converting the data to discrete. This results in a contingency table, which can easily be used in a chi-square test. This sort of test helps to determine which line is not performing like the others. Other options if converting to discrete include non-parametric hypothesis tests such as the Kruskal-Wallis test and the Mood's median test. Once this testing is done, the results must be interpreted with a focus on the p-value. This assists in determining whether or not the null hypothesis should be accepted. Once the statistical analysis is out of the way, other lean six sigma tools can then be used to increase a manufacturing system's profitability.

6.2.1.8. Monitoring the Waste Minimization Program Periodically

Once the waste minimization program has been performed and verified, it must be examined periodically to make sure that it is still implementing as planned and to make any necessary amendments. This entails monitoring the waste reduction amounts, operational outcome, and financial performance versus the objectives. New product or process variations should also be analyzed at the beginning to guarantee that the design minimizes environmental impact. This is simply achieved through inserting an additional waste minimization review related to the new product or process in the checklist or standard operating procedure.

Moreover, emphasis should be put on continuous improvement to enhance the current waste reduction program and to recognize new opportunities. It can be advantageous to perform periodic waste assessments, facility walk-throughs, and employee interviews with the original waste reduction team to achieve the defined objectives. When assessing the program, it is essential to consider the following points:

- Keeping track of program success and using recordered data from the past accomplishments
- Recognizing new ways of thinking about waste reduction
- Identifying zones needing improvement
- Documenting the results and finding in accordance with state and local regulations
- Clarifying the influence of new additions to the facility or program
- Keeping employees informed and interested

The waste minimization program is an ongoing process versus a one-time project. Normally, the first waste analysis and applied alternatives will aim at a reduction in the high-volume waste streams. Once these high-volume streams have been decreased, the team can concentrate on smaller-volume waste streams. The frequency at which the additional waste evaluations are conducted will rely on the budget of the company.

In general, manufacturing systems that conduct assessments one to four times annually have reached paybacks. Moreover, if there are specific conditions that are signs of a need for more review, additional waste assessments should be performed. These special conditions encompass:

- Changes in raw material or product requirements
- Elevated waste management costs
- New regulations
- New equipment, technology, and machines
- A key event with unfavorable environmental consequences (e.g., a major spill)

In order to be really efficient, an organizational culture of waste minimization should be promoted throughout the entire manufacturing system's processes by executive managers, engineers, and some selected staffs. This will make waste minimization a primary part of a manufacturing system's operation.

References

1. "Lean Manufacturing and the Environment," www.epa.gov/lean/environment/index.htm, retrieved January, 26, 2015.
2. Simpson, D., and Samson, D. (2010), "Environmental strategy and low waste operations: Exploring complementarities," *Business Strategy and the Environment* 19, no. 2, 104–118.
3. Karim, A., and Arif, U. Z. K. (2013), "A methodology for effective implementation of lean strategies and its performance evaluation in manufacturing organizations," *Business Process Management Journal* 19, no. 1, 169–196.
4. "Profitable Applications of Value Stream Mapping Tutorial," http://asq.org/learn-about-quality/lean/overview/value-stream-mapping.html, retrieved January 26, 2015.
5. Chiarini, A. (2014), "Sustainable manufacturing-greening processes using specific Lean Production tools: An empirical observation from European motorcycle component manufacturers," *Journal of Cleaner Production* 85, 226–233.
6. Elahi, B., Pakzad-Jafarabadi, Y., Etaati, L., and Seyedhosseini, S. M. (2011), "Optimization of supply chain planning with considering defective rates of products in each echelon," *Technology and Investment* 2, no. 3, 211–221.
7. Elahi, B., and Franchetti, M., "A New Optimization Model for Closed-Loop Supply Chain Networks," Technology Management Conference (ITMC), 2014 IEEE International (Chicago, June 12–15, 2014), 1–9.
8. Elahi, B., Etaati, L., and Seyed-Hosseini, S. M., "An Integrated Modelling for Supplier Selection and Optimal Lot Sizing: A Case Study of Four-Echelon Supply Chain," Technology Management Conference (ITMC), IEEE International (San Jose, CA, June 27–30, 2011), 877–884.
9. Akhavan, P., Elahi B., and Jafari, M. (2014), "A new integrated knowledge model in supplier selection: The case of an Asian automotive supply chain," *Journal of Education, Business and Society: Contemporary Middle Eastern Issues* 7, no. 4, 333–368.
10. Franchetti, M., *Solid Waste Analysis and Minimization: A Systems Approach* (New York: McGraw-Hill, 2009), 127–157.

Chapter 7

Hazardous Waste Management

Hazardous waste is a waste with properties that make it dangerous or potentially harmful to human health or the environment. Hazardous waste is generated from many sources, ranging from industrial manufacturing process wastes to batteries to fluorescent lightbulbs. Hazardous waste may come in many forms, including liquids, solids, gases, and sludge. In the United States, the treatment, storage, and disposal of hazardous waste are regulated under the Resource Conservation and Recovery Act (RCRA), passed by Congress in October 1976. The RCRA defines *hazardous waste* as: a solid waste, or combination of solid waste, which because of its quantity, concentration, or physical, chemical, or infectious characteristics may (a) cause, or significantly contribute to, an increase in mortality or an increase in serious irreversible, or incapacitating reversible, illness; or (b) pose a substantial present or potential hazard to human health or the environment when improperly treated, stored, transported, or disposed of, or otherwise managed.[1]

In order to develop and implement a regulatory framework capable of ensuring adequate protection, determining what is a hazardous waste is paramount; making this determination is a complex task that is a central component of the hazardous waste management regulations. The hazardous waste identification process, developed by the U.S. Environmental Protection Agency (EPA), is the crucial first step in the hazardous waste management system. Correctly determining whether a waste meets the RCRA definition of *hazardous waste* is essential to determining how the waste must be managed. The waste generator has the responsibility for determining if a waste is an RCRA hazardous waste.

7.1. Types of Hazardous Waste

Hazardous wastes are divided into listed wastes, characteristic wastes, universal wastes, and mixed wastes. Specific procedures determine how waste is identified, classified, listed, and delisted. Types of hazardous waste per the RCRA definition are explained in the sections that follow.[2]

7.1.1. Listed Wastes

By definition, the EPA determined that some specific wastes are hazardous. These wastes are incorporated into lists published by the agency. These lists are organized into three categories:

The F-list (nonspecific source wastes)—this list identifies wastes from common manufacturing and industrial processes, such as solvents that have been used in cleaning or degreasing operations. Because the processes producing these wastes can occur in different sectors of industry, the F-listed wastes are known as wastes from nonspecific sources.

The K-list (source-specific wastes)—this list includes certain wastes from specific industries, such as petroleum refining or pesticide manufacturing. Certain sludge and wastewaters from treatment and production processes in these industries are examples of source-specific wastes.

The P-list and the U-list (discarded commercial chemical products)—these lists include specific commercial chemical products in an unused form. Some pesticides and some pharmaceutical products become hazardous waste when discarded.

7.1.2. Characteristic Wastes

Waste that has not been specifically listed may still be considered a hazardous waste if it exhibits one of the four characteristics defined: ignitability, corrosivity, reactivity, or toxicity.

- **Ignitability:** Ignitable wastes can create fires under certain conditions, are spontaneously combustible, or have a flash point of less than 60°C (140°F). Examples include waste oils and used solvents. There are standard test methods recommended by the EPA that can be used to determine the ignitability of a specific waste.

- **Corrosivity:** Corrosive wastes are acids or bases (pH of less than or equal to 2, or greater than or equal to 12.5) that are capable of corroding metal containers, such as storage tanks, drums, and barrels. Battery acid is an example. There is one standard test method recommended by the EPA that may be used to determine corrosivity of a specific waste.

- **Reactivity:** Reactive wastes are unstable under "normal" conditions. They can cause explosions, toxic fumes, gases, or vapors when heated, compressed, or mixed with water. Examples include lithium-sulfur batteries and explosives. There are currently no standard test methods available to establish reactivity of a specific waste.
- **Toxicity:** Toxic wastes are harmful or fatal when ingested or absorbed (containing mercury, lead, etc.).[3] When toxic wastes are land disposed, contaminated liquid may leach from the waste and pollute groundwater. Toxicity is defined through a laboratory procedure called the Toxicity Characteristic Leaching Procedure (TCLP). The TCLP helps identify wastes likely to leach concentrations of contaminants that may be harmful to human health or the environment.

7.1.3. Universal Wastes

EPA's universal waste regulations streamline hazardous waste management standards for federally designated "universal wastes," which include batteries, pesticides, mercury-containing equipment, and bulbs (lamps). The regulations govern the collection and management of these widely generated wastes, thus facilitating environmentally sound collection and proper recycling or treatment. These regulations also ease the regulatory burden on retail stores and others that wish to collect these wastes and encourage the development of municipal and commercial programs to reduce the quantity of these wastes going to municipal solid waste landfills or combustors. In addition, the regulations also ensure that the wastes subject to this system will go to appropriate treatment or recycling facilities pursuant to the full hazardous waste regulatory controls.

7.1.4. Mixed Wastes

Mixed waste contains both radioactive and hazardous waste components. As a result, both treatment and regulation are complex. Mixed wastes are regulated by the RCRA and the Atomic Energy Act (AEA) of 1954. The U.S. Nuclear Regulatory Commission and the U.S. Department of Energy regulate the radioactive portion of mixed waste under AEA authority, while the EPA regulates the hazardous waste portion of mixed waste under RCRA authority. Most commercially generated mixed waste is classified as low-level mixed waste (LLMW). LLMW is waste that contains low-level radioactive waste and hazardous waste. LLMW is generated commercially in all 50 states at industrial, hospital, and nuclear power plant facilities in a number of processes, such as medical diagnostic testing and research, pharmaceutical and biotechnology development, pesticide research, and nuclear power plant operations. Other categories are high-level mixed waste (HLW) resulting from reprocessing spent nuclear fuel and irradiated targets from reactors, and mixed transuranic waste (MTRU) resulting from nuclear weapons fabrication, plutonium-bearing reactor fuel fabrication, and spent fuel reprocessing.

7.2. Hazardous Waste Identification

Proper hazardous waste identification is essential to the success of the RCRA program. The hazardous waste identification (HWID) process can be a very complex task, and it is best to approach the issue by asking a series of questions in a stepwise manner. The HWID process consists of four questions: Is the material a solid waste? Is the waste specifically excluded from the RCRA? Is the waste a listed hazardous waste? Does the waste exhibit a characteristic of hazardous waste?

The RCRA Subtitle C program uses the term *solid waste* to denote something that is a waste. In order for a material to be classified as a hazardous waste, it must first be a solid waste. Only a small fraction of all RCRA solid wastes actually qualify as hazardous wastes. Distinguishing between hazardous and nonhazardous wastes is not a simple matter of chemical and toxicological analysis; other factors must be considered before evaluating the actual hazard that a waste's chemical composition poses. Regulation of certain wastes may be impractical, unfair, or otherwise undesirable, regardless of the hazards they pose. For instance, household waste can contain dangerous chemicals, like solvents and pesticides, but making households subject to the strict RCRA waste management regulations would create a number of practical problems. Regulations exempted or excluded certain wastes, such as household wastes, from the hazardous waste definition and regulations. Determining whether a waste is excluded or exempted from hazardous waste regulation is the second step in the HWID process. Only after determining that a solid waste is not somehow excluded from hazardous waste regulation should the analysis proceed to evaluate the actual chemical hazard that a waste poses.

A hazardous waste listing is a narrative description of a specific type of waste that the EPA considers dangerous enough to warrant regulation. Hazardous waste listings describe wastes from various industrial processes, wastes from specific sectors of industry, and wastes in the form of specific chemical formulations. Before developing a hazardous waste listing, the EPA thoroughly studies a particular waste stream and the threat it can pose to human health and the environment. If the waste poses enough of a threat, the EPA includes a precise description of that waste on one of the hazardous waste lists in the regulations. Thereafter, any waste fitting that narrative listing description is considered hazardous, regardless of its chemical composition or any other potential variable.

The final steps in the HWID process determine whether a waste actually poses a sufficient chemical or physical hazard to merit regulation. A hazardous waste characteristic is a property that when present in a waste indicates the waste poses a sufficient threat to merit regulation as hazardous. When defining hazardous waste characteristics, the EPA identifies, where practicable, analytical tests capable of detecting or demonstrating the presence of the characteristic. Whether a waste displays a hazardous characteristic generally depends on how it fares in one of the characteristics tests; therefore, the chemical makeup or other factors about the composition of a particular waste typically determine whether it tests as hazardous for a characteristic.

7.3. Hazardous Waste Generators and Transporters

A hazardous waste generator is any person or site whose processes and actions create hazardous waste.[4] Generators are divided into three categories based on the quantity of waste they produce. Each class of generator must comply with its own set of requirements. Additionally, most states are authorized to implement the RCRA program. States can be authorized for categories of hazardous waste generators that are different from the federal categories.

Large-quantity generators (LQGs) generate 1,000 kilograms (2,200 pounds) per month or more of hazardous waste, more than 1 kilogram (2.2 pounds) per month of acutely hazardous waste, or more than 100 kilograms (220 pounds) per month of acute spill residue or soil. Requirements for LQGs include:

- LQGs may accumulate waste on-site for only 90 days. Certain exceptions apply.
- LQGs do not have a limit on the amount of hazardous waste accumulated on-site.
- There must always be at least one employee available to respond to an emergency. This employee is the emergency coordinator responsible for coordinating all emergency response measures. LQGs must have detailed, written contingency plans for handling emergencies.

- LQGs must submit a biennial hazardous waste report.

Small-quantity generators (SQGs) generate more than 100 kilograms, but less than 1,000 kilograms, of hazardous waste per month. Requirements for SQGs include:

- SQGs may accumulate hazardous waste on-site for 180 days without a permit (or 270 days if shipping a distance greater than 322 kilometers [200 miles]).
- The quantity of hazardous on-site waste must never exceed 6,000 kilograms (13,228 pounds).
- There must always be at least one employee available to respond to an emergency. This employee is the emergency coordinator responsible for coordinating all emergency response measures. SQGs are not required to have detailed, written contingency plans.

Conditionally exempt small-quantity generators (CESQGs) generate 100 kilograms or less per month of hazardous waste or 1 kilograms or less per month of acutely hazardous waste. Requirements for CESQGs include:

- CESQGs must identify all the hazardous waste generated.
- CESQGs may not accumulate more than 1,000 kilograms of hazardous waste at any time.
- CESQGs must ensure that hazardous waste is delivered to a person or facility that is authorized to manage it.

Hazardous waste transporters are individuals or entities that move hazardous waste from one site to another by highway, rail, water, or air. This includes transporting hazardous waste from a generator's site to a facility that can recycle, treat, store, or dispose of the waste. It can also include transporting treated hazardous waste to a site for further treatment or disposal. Requirements for transporters include:

- Obtaining an EPA identification number
- Complying with the manifest system
- Responding appropriately to hazardous waste discharges
- Complying with all applicable DOT regulations

The Hazardous Waste Manifest System is a set of forms, reports, and procedures that track hazardous waste from the time it leaves the generator facility until it reaches the waste management facility that will store, treat, or dispose of the hazardous waste. The system allows the waste generator to verify that its waste has been properly delivered and that no waste has been lost or unaccounted for in the process. The key component of this system is the Uniform Hazardous Waste Manifest which is a form prepared by all generators who transport, or offer for transport, hazardous waste for off-site treatment, recycling, storage, or disposal. The manifest is required by both the Department of Transportation and the EPA.

7.4. Hazardous Waste Treatment, Storage, and Disposal

The Resource Conservation and Recovery Act creates a cradle-to-grave management system for hazardous waste to ensure proper treatment, storage, and disposal in a manner protective of human health and the environment. The RCRA hazardous waste permitting program helps ensure the safe treatment, storage, and disposal of hazardous wastes by establishing specific requirements that must be followed when managing those wastes. Permits for the treatment, storage, or disposal of hazardous wastes are issued by authorized states or by EPA regional offices. Under the RCRA, regulations have been established for designing and operating requirements for land disposal units (LDUs). The requirements are intended to minimize pollution resulting from the disposal of hazardous waste in or on the land. Types of LDUs for hazardous waste disposal include landfill, surface impoundment, waste pile, land treatment unit, injection well, salt dome formation, salt bed formation, underground mine, and underground cave.

Storage is the holding of waste for a temporary period of time prior to the waste being treated, disposed of, or stored elsewhere. Hazardous waste is commonly stored prior to treatment or disposal, and it must be stored in containers, tanks, containment buildings, drip pads, waste piles, or surface impoundments that comply with RCRA regulations. Permitted storage units include:

- **Containers:** A hazardous waste container is any portable device in which a hazardous waste is stored, transported, treated, disposed of, or otherwise handled. The most common hazardous waste container is the 55-gallon (about 208-liter) drum. Other examples of containers are tanker trucks, railroad cars, buckets, bags, and even test tubes.

- **Tanks:** Tanks are stationary devices constructed of non-earthen materials used to store or treat hazardous waste. Tanks can be open-topped or completely enclosed and are constructed of a wide variety of materials, including steel, plastic, fiberglass, and concrete.
- **Containment buildings:** Containment buildings are completely enclosed, self-supporting structures (i.e., they have four walls, a roof, and a floor) used to store or treat non-containerized hazardous waste.
- **Drip pads:** A drip pad is a wood drying structure used by the pressure-treated wood industry to collect excess wood preservative drippage. Drip pads are constructed of non-earthen materials with a curbed, free-draining base that is designed to convey wood preservative drippage to a collection system for proper management.
- **Waste Piles.** A waste pile is an open, uncontained pile used for treating or storing waste. Hazardous waste piles must be placed on top of a double liner system to ensure leachate from the waste does not contaminate surface or groundwater supplies.
- **Surface Impoundments:** A surface impoundment is a natural topographical depression, man-made excavation, or diked area such as a holding pond, storage pit, or settling lagoon. Surface impoundments are formed primarily of earthen materials and are lined with synthetic plastic liners to prevent liquids from escaping.

7.5. Hazardous Waste Recycling

The EPA developed the hazardous waste recycling regulations to promote the reuse and reclamation of useful materials in a manner that is safe and protective of human health and the environment. Many industrial hazardous wastes can be recycled safely and effectively. A hazardous waste is recycled if it is used, reused, or reclaimed. Furthermore, RCRA hazardous waste regulation makes an important distinction between materials that are used or reused without reclamation and those that must be reclaimed before reuse. A material is reclaimed if it is processed to recover a usable product or if it is regenerated. Common hazardous waste reclamation activities involve recovery of spent solvents (e.g., recovery of acetone) or metals (e.g., recovery of lead). Hazardous waste recycling activities include combustion for energy recovery; use constituting disposal; reclamation; and direct use and reuse. Recycling hazardous waste fulfills two of the RCRA's goals by reducing the consumption of raw materials and energy and by reducing the volume of waste materials that must be treated and disposed of.

Combustion for energy recovery involves burning the hazardous waste directly as a fuel or using it as an ingredient to produce a fuel. Used solvents, for example, are frequently burned to produce heat or generate electricity. Because of the potential for release of harmful constituents from burning these wastes, the EPA regulates this recycling activity as strictly as any other type of hazardous waste combustion. The EPA requires combustion units that burn hazardous waste for energy recovery to obtain a permit and meet certain performance and operating standards under the boiler and industrial furnace regulations.

Use constituting disposal involves applying a hazardous waste directly to the land or incorporating it into a product that will be applied to the land. Examples include using hazardous waste as a fertilizer or as an ingredient in asphalt. The EPA strictly regulates land disposal of hazardous waste because of the potential for soil and groundwater contamination. Recycling a hazardous waste in a manner that constitutes disposal (land application) presents similar risks. The harmful constituents in hazardous wastes must be treated to reduce their toxicity and ability to leach into soil and groundwater before the wastes are applied to the land.

Reclamation is processing a material to recover a usable product, such as recovering mercury from broken thermometers or regenerating a material, such as cleaning used solvents to make them pure again. Reclamation activities are regulated differently depending on the type of hazardous waste to be recycled. Certain reclaimed materials enjoy "relief" from all hazardous waste regulations. Other materials, however, are subject to full regulation when reclaimed. The EPA made this distinction based on the level of threat posed by common industry practices associated with reclaiming different types of materials.

The final type of hazardous waste recycling activity is using a waste directly (without reclamation) as an ingredient in an industrial process to make a product or using a waste directly as a substitute for a product. Under this activity, a facility will use a hazardous waste directly in place of a product if the waste is similar enough to function in a similar manner. Since direct reuse of the material presents a low risk to human health and the environment, the EPA does not regulate these activities unless the waste will be burned or placed on the land. The agency will evaluate the legitimacy of a recycling practice by ensuring that it is not an attempt to avoid proper treatment or disposal and that the material is recycled in a timely manner.

References

1. U.S. Code 42 § 6903, www.gpo.gov/fdsys/pkg/USCODE-2010-title42/pdf/USCODE-2010-title42-chap82-subchapI-sec6903.pdf, retrieved June 2015.

2. "Hazardous Waste," U.S. Environmental Protection Agency,www.epa.gov/epawaste/hazard/, retrieved September 2014.

3. "Characteristic Wastes," U.S. Environmental Protection Agency,www.epa.gov/osw/hazard/wastetypes/characteristic.htm, retrieved June 2015.

4. "Hazardous Waste Generators," U.S. Environmental Protection Agency,www2.epa.gov/hwgenerators, retrieved June 2015.

Chapter 8

Energy Resource Management and Energy Efficiency

Management of energy resources includes the production, distribution, and consumption of various forms of energy for human use in lighting, heating, cooling, industrial manufacturing, water treatment and distribution, food production and processing, transportation, and many others. The Association of German Engineers (VDI) provides a related definition: "Energy management is the proactive, organized and systematic coordination of procurement, conversion, distribution and use of energy to meet the requirements, taking into account environmental and economic objectives."[1]The primary environmental objectives are resource conservation, prevention of environmental degradation, and global climate protection. Energy production mainly from fossil fuel resources has major consequences on natural resource depletion, water and land resources, and global warming.

According to the U.S. Environmental Protection Agency (EPA), production and use of energy (most of which is from fossil fuels) accounts for more than 80% of U.S. greenhouse gas emissions, contributing to global warming.[2] The environmental impact of electricity generation is significant because modern society uses large amounts of electrical power. Power plants can require large amounts of water for cooling, and on average, a kilowatt hour of electricity requires 95 liters (25 gal) of water to be withdrawn from rivers or lakes.

Energy costs are a major component in most business budgets, and effective energy management is one of the most cost-effective strategies a company can use to manage rising energy costs and growing concerns over greenhouse gas emissions. Energy should be regarded as a business cost, like raw material or labor, and reducing cost is the most compelling business reason for improving energy management, followed by reducing risk of energy price fluctuations. Introducing strong and robust energy management systems and ways to identify performance improvements can drive significant cost savings and competitive advantages and can mitigate against energy price volatility. Companies that have an in-depth knowledge and understanding of their energy use and systems to manage it have demonstrated increased productivity, better staff engagement, and reputational benefits.

Two most basic elements of sustainable energy management are energy conservation and sustainable energy sources. Energy conservation refers to efforts made to reduce energy consumption. Energy conservation can be achieved through increased efficient energy use, in conjunction with decreased energy consumption and/or reduced consumption from conventional energy sources. Sustainable energy sources are most often regarded as including all renewable energy sources, such as hydroelectricity, solar energy, wind energy, wave power, geothermal energy, bioenergy, and tidal power. Sustainable energy usually also includes technologies that improve energy efficiency.

8.1. Organizational Structure for Energy Management

For effective energy management at any business, and particularly for large or energy-intensive businesses, it is important to integrate management activities into the organizational structure and to establish a separate organizational unit to implement energy management. The placement of responsibility for energy management within the organizational structure is an important decision. Energy management has impacts on the whole organization, and to be effective, all parts of the organization need to be accessed. Where energy management should be located depends on the structure of the organization and its maturity in terms of energy management practices. Some points to consider are:

- It should be determined whether responsibility should be concentrated or distributed.
- Energy management is more than a technical function—it is management too.
- All managers are responsible in some way for energy.
- Accountability for energy use should be distributed to those who control it.

The critical issue in organizing for energy management is its integration into the overall management structure and process. While leadership may be placed in the hands of an individual or group of individuals, just like other management functions, energy management needs to be incorporated into the roles and responsibilities of all line managers. The energy management function, whether vested in one energy manager or distributed among a number of middle managers, usually resides somewhere in the organization between senior management and those who control the end use of energy. Exactly how and where that function is placed is a decision that needs to be made in view of the existing corporate structure. Saving energy has tended to be seen as a technical activity, and the energy manager is often located in a technical section within the organization. This may be a good base for gaining control in the initial phase of an energy management program, but it could prove to be less appropriate for training or energy information activities. It is important to integrate the functions of an energy manager with those of human resources for the motivation and training activities, as well as those of the finance division to operate the financial control and accounting procedures required in later project implementation phases.

Another option is to employ outside consultants who can provide wide experience and expertise. This may be the best option in technical situations when consultants can be used to support internal personnel, but it lacks the network of relationships and day-to-day contact that is crucial for informing and motivating staff.

In practice, there may be no single ideal home for all energy management activities, and the optimum location may need to reflect this, altering over time as the organization moves from one phase of its energy management program to the next. Each option has its own advantages and shortcomings. Important questions concerning the concentration of the energy management function are:

- Should all energy staff be kept together in a combined unit?
- Or is it more appropriate for them to be dispersed across the organization?

Which of these options will prove to be best, not just in the short term but also in the long run, will depend on the organization's specific circumstances. If energy management is based in a technical section, there is a danger that saving energy may be marginalized as a specialized technical activity. Energy is an organization-wide management issue, not a technical specialty. It is essential that:

- All managers understand that controlling energy consumption is one of their managerial responsibilities.
- They accept and act on this "new" understanding and are made accountable for their own energy consumption.

8.2. Systems Approach and Strategies for Organizational Energy Management

A systematic approach to energy management can assist an organization in improving its energy and financial performance while distinguishing it as an environmental leader. The central task of energy management is to reduce energy consumption and costs for the provision of energy in buildings and facilities without compromising work processes. Especially the availability and service life of the equipment and the ease of use should remain the same.

According to the ENERGY STAR, a voluntary program of the U.S. EPA that assists businesses and individuals with superior energy management, the following are important steps in establishing an organizational energy management program and implementing improvements:[3]

Step 1: Make Commitment

Step 2: Assess Performance

Step 3: Set Goals

Step 4: Create Action Plan

Step 5: Implement Action Plan

Step 6: Evaluate Progress

Step 7: Recognize Achievements

The sequence and integrated relationship between the above steps are depicted in Figure 8.1.

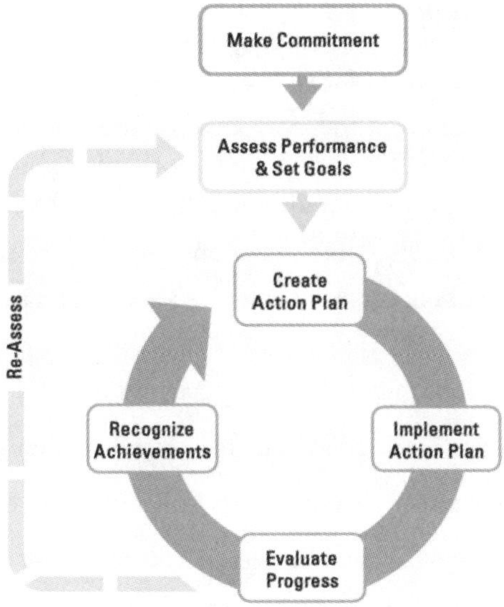

Figure 8.1. Steps for Organizational Energy Management
Source: U.S. Environmental Protection Agency, ENERGY STAR.

Step 1: Commit to Continuous Improvement

Organizations seeing the financial returns from superior energy management continuously strive to improve their energy performance. Their success is based on regularly assessing energy performance and implementing steps to increase energy efficiency. No matter the size or type of organization, the common element of successful energy management is commitment. Organizations make a commitment to allocate staff and funding to achieve continuous improvement. To establish their energy program, leading organizations form a dedicated energy team and institute an energy policy.

Step 1.1: Appoint an Energy Director. Sets goals, tracks progress, and promotes the energy management program.

Step 1.2: Establish an Energy Team. Executes energy management activities across different parts of the organization and ensures integration of best practices.

Step 1.3: Institute an Energy Policy. Provides the foundation for setting performance goals and integrating energy management.

Step 2: Assess Performance

Understanding current and past energy use is how many organizations identify opportunities to improve energy performance and gain financial benefits. Assessing performance is the periodic process of evaluating energy use for all major facilities and functions in the organization and establishing a baseline for measuring future results of efficiency efforts. Assessing the energy performance helps an organization to:

- Categorize current energy use by fuel type, operating division, facility, product line, and so forth
- Identify high performing facilities for recognition and replicable practices
- Prioritize poor performing facilities for immediate improvement
- Understand the contribution of energy expenditures to operating costs
- Develop a historical perspective and context for future actions and decisions
- Establish reference points for measuring and rewarding good performance

Key aspects include:

Step 2.1: Gather and track data. Collect energy use information and document data over time.

Step 2.2: Establish baselines. Determine the starting point from which to measure progress.

Step 2.3: Benchmark. Compare the energy performance of facilities to each other, peers and competitors, and over time to prioritize which facilities to focus on for improvements.

Step 2.4: Analyze. Understand the energy use patterns and trends.

Step 2.5: Technical assessments and audits. Evaluate the operating performance of facility systems and equipment to determine improvement potential.

Step 3: Set Goals

Performance goals drive energy management activities and promote continuous improvement. Setting clear and measurable goals is critical for understanding intended results, developing effective strategies, and reaping financial gains. Well-stated goals guide daily decision making and are the basis for tracking and measuring progress. Communicating and posting goals can motivate staff to support energy management efforts throughout the organization. The Energy Director in conjunction with the Energy Team typically develops goals. Setting goals helps with the following:

- Setting the tone for improvement throughout the organization
- Measuring the success of the energy management program
- Helping the Energy Team to identify progress and setbacks at a facility level
- Fostering ownership of energy management, creating a sense of purpose, and motivating staff
- Demonstrating commitment to reducing environmental impacts
- Creating schedules for upgrade activities and identify milestones

To develop effective performance goals:

Step 3.1: Determine scope. Identify organizational and time parameters for goals.

Step 3.2: Estimate potential for improvement. Review baselines, benchmark to determine the potential and order of upgrades, and conduct technical assessments and audits.

Step 3.3: Establish goals. Create and express clear, measurable goals, with target dates, for the entire organization, facilities, and other units.

Step 4: Create Action Plan

With goals in place, an organization is poised to develop a road map to improve energy performance. Successful organizations use a detailed action plan to ensure a systematic process for implementing energy performance measures. Unlike the energy policy, the action plan is regularly updated, most often on an annual basis, to reflect recent achievements, changes in performance, and shifting priorities. Creating an inclusive strategy that establishes roles and actions throughout the organization can help to integrate good energy management practices. When developing an action plan, the following could be considered:

- Brainstorming with various departments to identify ways they can contribute
- Holding a competition to seek ideas for energy efficiency from across the organization
- Gathering recommendations from the Energy Team and other key personnel

While the scope and scale of the action plan are often dependent on the organization, the steps below outline a basic starting point for creating a plan.

Step 4.1: Define technical steps and targets. Evaluate technical assessments and audit results, identify gaps between current performance and goals, and identify the steps necessary for upgrading and moving facilities from current performance to the desired level of performance as defined by the goals.

Step 4.2: Determine roles and resources. Determine who should be involved within the organization and what their responsibilities will be; determine the degree to which consultants, service providers, vendors, and other product providers will be used. For each project or program in the action plan, estimate the cost for each item in terms of both human resources and capital/expense outlay. Develop the business case for justifying and gaining funding approval for action plan projects and resources need.

Step 5: Implement Action Plan

Gaining the support and cooperation of key people at different levels within the organization is an important factor for successful action plan implementation in many organizations. In addition, reaching energy management goals frequently depends on the awareness, commitment, and capability of the people who will implement the projects. To implement an action plan, the following steps can be considered:

Step 5.1: Create a communication plan. Develop targeted information for key audiences about the energy management program.

Step 5.2: Raise awareness. Build support at all levels of the organization for energy management initiatives and goals.

Step 5.3: Build capacity. Through training, access to information, and transfer of successful practices, procedures, and technologies, capacity of the staff can be extended.

Step 5.4: Motivate. Create incentives that encourage staff to improve energy performance to achieve goals.

Step 5.5: Track and monitor. Use the tracking system developed as part of the action plan to track and monitor progress regularly.

Step 6: Evaluate Progress

Evaluating progress includes formal review of both energy use data and the activities carried out as part of the action plan as compared to the performance goals. Evaluation results and information gathered during the formal review process are used by many organizations to create new action plans, identify best practices, and set new performance goals.

Regular evaluation of energy performance and the effectiveness of energy management initiatives also allows energy managers to:

- Measure the effectiveness of projects and programs implemented
- Make informed decisions about future energy projects
- Reward individuals and teams for accomplishments
- Document additional savings opportunities as well as non-quantifiable benefits that can be leveraged for future initiatives

Key steps involved include:

Step 6.1: Measure results. Compare current performance to established goals.

Step 6.2: Review action plan. Understand what worked well and what did not in order to identify best practices.

Step 7: Recognize Achievements

Providing and seeking recognition for energy management achievements is a proven step for sustaining momentum and support for the program. Providing recognition to those who helped the organization achieve these results motivates staff and employees and brings positive exposure to the energy management program. Receiving recognition from outside sources validates the importance of the energy management program to both internal and external stakeholders and offers positive exposure for the organization as a whole.

Key steps in providing and gaining recognition include:

Step 7.1: Providing internal recognition. This applies to individuals, teams, and facilities within the organization.

Step 7.2: Receiving external recognition. Significant recognition comes from government agencies, the media, and other third-party organizations that reward achievement.

8.3. Evaluation of Energy Consumption

Energy monitoring is primarily a management technique that uses energy information as a basis to eliminate waste, reduce and control current level of energy use, and improve the existing operating procedures. It builds on the principle that one cannot manage what one cannot measure. Evaluating energy performance requires good information on how, when, and where energy is being used. Collecting and tracking this information are necessary for establishing baselines and managing energy use. Monitoring provides energy managers with feedback on operating practices, results of energy management projects, and guidance on the level of energy use that is expected in a certain period or for a certain activity. It is important to note that monitoring also gives early warning of unexpected excess consumption caused by equipment malfunction, operator error, unwanted user behavior and lack of effective maintenance.

Some of the goals of energy monitoring are to determine the following:

- Whether current energy usage is better or worse than before
- Trends in energy consumption that reflect seasonal, weekly, and other operational parameters
- How much future energy use is likely to vary if aspects of the business change
- Specific areas of wasted energy
- Comparison with other businesses with similar characteristics (bench marking)
- How much a business has reacted to changes in the past
- How to develop performance targets for an energy management program

The essential elements of an energy-monitoring program are:[4]

- **Recording.** Measuring and recording energy consumption.
- **Analyzing.** Correlating energy consumption to a measured output, such as production quantity.
- **Comparing.** Comparing energy consumption to an appropriate standard or benchmark.
- **Setting Targets.** Setting targets to reduce or control energy consumption.
- **Monitoring.** Comparing energy consumption to the set target on a regular basis.
- **Reporting.** Reporting the results including any variances from the targets that have been set.
- **Controlling.** Implementing management measures to correct any variances that may have occurred.
- **Checking** the accuracy of energy invoices
- **Allocating** energy costs to specific departments
- **Determining** energy performance/efficiency
- **Recording** energy use, so that projects intended to improve energy efficiency can be checked
- **Highlighting** performance problems in equipment or systems

Organizations of all sizes have some form of established systems for gathering and tracking energy use data, though the level and scope of data collection may vary from organization to organization. Some may choose to collect data from sub-meters on individual processes, while others may only look at a utility bill. It is important to inventory all energy purchased and generated on-site (electricity, gas, steam, waste fuels) in physical units (kilowatt hours, Mcfs, MMBTUs, pounds of steam, etc.) and on a cost basis. Energy data may also reside in the accounting department, may be held centrally or at each facility, or can be acquired by contacting the appropriate utilities or energy service providers. It is advisable to gather at least two years of monthly data or a more frequent interval if available. A system for tracking energy usage can range from a simple spreadsheet to detailed databases and information technology (IT) systems.

Information related to energy use at industrial plants can be obtained from the following sources:

- Plant level information can be derived from financial accounting systems or utilities cost center.
- Plant department-level information can be found in comparative energy consumption data for a group of similar facilities, service entrance meter readings, and so forth.
- System-level (for example, boiler plant) performance data can be determined from sub-metering data.
- Equipment-level information can be obtained from nameplate data, run-time and schedule information, and sub-metered data on specific energy-consuming equipment.

After collection of energy-consumption, energy-cost, and energy-production data, the next stage of the monitoring process is to study and analyze the data to understand what is happening in the plant. It is strongly recommended that the data be presented graphically. A better appreciation of variations is almost always obtained from a visual presentation, rather than from a table of numbers. Graphs generally provide an effective means of developing the energy production relationships, which explain what is going on in the plant.

8.4. Industrial Energy Management

According to the U.S. Energy Information Administration, the industrial sector accounts for about one-third of the total energy used in the United States and efficient energy management in the industrial sector is critical for sustainability and environmental protection, apart from cost savings and financial benefits. Opportunities for industrial energy efficiency improvement and energy savings typically can be found in four areas:

- Operational, such as eliminating unnecessary use of existing equipment
- Small capital projects, such as lighting upgrades
- Large capital projects, such as building renovations and process equipment upgrades
- Procurement, such as renegotiating utility supply contracts

Energy audits assist industrial companies or facilities in understanding how they use energy and help to identify the areas where waste occurs and where opportunities for improvement exist. The type of industrial energy audit conducted depends on the function, size, and type of the industry; the depth to which the audit is needed; and the potential and magnitude of energy savings and cost reduction desired. Based on these criteria, an industrial energy audit can be classified into two types: a preliminary audit (walk-through audit) and a detailed audit (diagnostic audit).

In a preliminary energy audit, readily available data are mostly used for a simple analysis of energy use and performance of the plant. This type of audit does not require a lot of measurement and data collection. These audits take a relatively short time, and the results are more general, providing common opportunities for energy efficiency. The economic analysis is typically limited to calculation of the simple payback period, or the time required paying back the initial capital investment through realized energy savings.

For detailed (or diagnostic) energy audits, more thorough data and information are required. Measurements and a data inventory are usually conducted, and different energy systems (pump, fan, compressed air, steam, process heating, etc.) are assessed in detail. Hence, the time required for this type of audit is longer than that of preliminary audits. The results of these audits are more comprehensive and useful, since they give a more accurate picture of the energy performance of the plant and more specific recommendation for improvements. The economic analysis conducted for the efficiency measures recommended typically go beyond the simple payback period and usually include the calculation of an internal rate of return (IRR), net present value (NPV), and often also life cycle cost (LCC).

Overall, there are three main steps (excluding the post-audit activities) each of which has several sub-steps. These three main steps are energy audit preparation, execution, and reporting. Before starting the energy audit, the criteria against which the audit will be conducted should be defined. Typical criteria that should be taken into consideration include audit objective, audit type, audit methodology and standards, staff involvement, site or utility boundary, timeline, and reporting requirements. Some of the most relevant steps for the audit process are as follows:

- **Defining the audit scope:** The audit scope needs to consider the available resources, such as staff, time, audit boundaries, level of analysis, expected results, degree of detail, and budget for conducting the energy audit.

- **Selection of energy audit team:** The selection of the energy auditing team should consider if the audit will be conducted by the plant's internal staff or if an outside consultant should be hired.
- **Making an audit plan:** An audit plan outlines the audit strategy and procedure and should provide scope of the audit, time of the audit, and its duration as well as the timeline for each step of the audit process, elements of the audit that have a high priority, responsibilities and tasks of each audit team member, and format of the audit report and its outline.
- **Preparing an audit checklist:** The audit checklist helps the auditor to conduct the work in a systematic and consistent way and should include steps to be taken during the energy audit, data and information that should be collected, existing measurement instrument and the data recorded, required measurements during the energy audit and the list of parameters to be measured, major equipment to be assessed in more detail, list of main components of the results section of the audit report, and any other major concerns and considerations.
- **Conducting the initial walk-through visit:** The purpose of the initial walk-through visit is for the energy audit team to become familiar with the facility to be audited, including processes and utilities to be audited in detail later and existing measurement instrumentation on the equipment and the data recorded, so that extra measurement and data collection requirements during the audit can be determined.
- **Collecting energy bills and available data and information:** Energy bills along with other current and historical energy- and production-related data and information should be collected at the beginning of the audit process, including energy bills and invoices, production data, climatic data as needed, architectural and engineering plans of the plant and its equipment, general information about the plant (year of construction, ownership status, renovations, types of products, operation schedule, operating hours, scheduled shut-downs, etc.), and status of energy management and any energy-saving measures implemented.
- **Conducting the preliminary analysis:** The preliminary analysis helps the energy auditor to better understand the plant by providing a general picture of the plant energy use, operation, and energy losses. A flowchart can be constructed that shows the energy flows of the system being audited. An overview of unit operations, important process steps, areas of material and energy use, and sources of waste generation should be presented in this flowchart.

- **Analyzing energy bills:** Energy bills, especially those for electricity and natural gas, are very useful for understanding and analyzing a plant's energy costs. It is important to understand the different components of these bills so that a correct and helpful analysis can be conducted.
- **Inventory and measurement of energy use:** Gathering data through an inventory and measurement is one of the main activities of energy auditing. Some data are readily available and can be collected from different divisions of the plant being audited. Some other data can be collected through measurement and recording. The most common data measured during the auditing process are liquid and gas fuel flows; electrical measurements, such as the voltage, current intensity, and power as well as power factor; temperatures of solid and liquid surfaces; pressure of fluids in pipes, furnaces, or vessels; exhaust gas emissions (CO_2, CO, O_2, and smoke); and relative humidity and luminance levels.
- **Benchmarking and comparative energy performance analysis:** Energy efficiency benchmarking and comparisons can be used to assess a company's performance relative to that of its competitors or its own performance in the past. Benchmarking can also be used for assessing the energy performance improvement achieved by the implementation of energy efficiency measures.
- **Identifying energy efficiency and energy cost-reduction opportunities:** Some of the common industrial systems that are typically considered for such assessments are compressed air, motors, process heating and cooling, and pumping and steam systems. A variety of resources and tools are available from sources such as the U.S. Department of Energy, U.S. EPA ENERGY STAR, and others for evaluation of the actual performance of such systems and equipment against the designed performance level or best available technology and practices.
- **Benefit-cost analysis of energy efficiency opportunities:** After identifying the list of energy efficiency measures applicable to the facility, the auditor can also conduct an economic feasibility analysis, a so-called benefit-cost analysis, for the measures and make recommendations for their implementation. LCC, discounting, NPV, IRR, savings-to-investment ratio, and payback periods are the common economic analyses for the assessment of financial viability of the energy efficiency measures.
- **Preparing an energy audit report:** After finishing the energy audit, the audit team should write an energy audit report. In the report, the auditors should explain their work and the results in a well-structured format. The energy audit report should be concise and precise and should be written in a way that is easy for the target audience to comprehend.

- **Post-audit activities:** In practice there are often barriers that prevent the successful implementation of energy efficiency measures recommended in an energy audit report. Therefore, it is helpful to establish a clear procedure to ensure the successful realization of recommended improvements. An implementation action plan should be described in a simple way with clear goals, saving targets, and definitions of roles and responsibilities for its execution.

References:

1. "Energy Management Approach," US Green Energy Technologies, www.usgreensolar.com/energy_management.asp, retrieved June 2015.

2. "Climate Impacts on Energy," U.S. Environmental Protection Agency, www.epa.gov/climatechange/impacts-adaptation/energy.html, retrieved June 2015.

3. "The ENERGY STAR Guidelines for Energy Management," U.S. Environmental Protection Agency, www.energystar.gov/buildings/tools-and-resources/energy-star-guidelines-energy-management, retrieved September 2014; Figure 8.1,www.energystar.gov/buildings/about-us/how-can-we-help-you/build-energy-program/guidelines, retrieved June 2015.

4. "Energy Monitoring & Targeting," Bureau of Energy Efficiency, Government of India, www.beeindia.in/energy_managers_auditors/documents/guide_books, retrieved June 2015.

Chapter 9

Sustainable Energy and Sustainable Energy Sources

Sustainable energy is the form of energy obtained from nonexhaustible resources such that the provision of this form of energy serves the needs of the present without compromising the ability of future generations to meet their needs. Generally, *sustainable energy* refers to an energy source that is an alternative to conventional energy sources, most commonly fossil fuels like coal, petroleum, and natural gas. These alternatives are intended to address concerns on the undesirable consequences of conventional energy usage on the environment and sustainability. Technologies that promote sustainable energy include renewable energy sources, such as hydroelectricity, solar energy, wind energy, wave power, geothermal energy, bioenergy and biofuels, tidal power, and also technologies designed to improve energy efficiency.

There have been extensive technological advancements in the field of sustainable energy in recent decades and further research, development, and commercialization are in progress. Costs of these sustainable energy technologies have also fallen dramatically in recent years, and continue to fall. Most of these technologies are either economically competitive to conventional energy technologies or close to being so, and with increasing government policies and public support in favor of sustainable energy, these markets are expanding. Many of the sustainable energy systems are available today in large utility-scale as well as smaller industrial-scale, commercial-scale, and residential-scale.

Figure 9.1 shows that shares of renewable energy in total U.S. energy production and consumption in 2013 have been 11.2% and 9.4%, respectively. Figure 9.2 shows the contribution of the different renewable energy technologies in the United States, with biomass energy as the leading contributor, followed by hydropower. Figure 9.3 shows the top U.S. states for

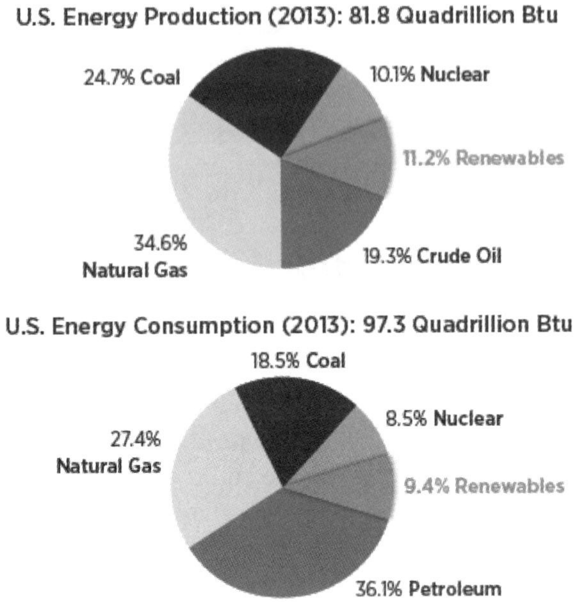

Figure 9.1. U.S. Energy Production and Consumption Including Renewables in 2013. ***Source: U.S. Department of Energy, 2014.***

installed electricity generation capacity in 2013 with various renewable energy technologies.[1]

9.1. Benefits of Sustainable Energy

Sustainable energy sources that are renewable, including solar energy, wind energy, biomass energy and biofuels, geothermal energy, and hydro-source energy, provide substantial benefits for the climate, human health, and the economy. The list of benefits include:

- Little to no global warming emissions
- Improved public health and environmental quality
- A vast and inexhaustible energy supply
- Jobs and other economic benefits
- Stable energy prices
- A more reliable and resilient energy system

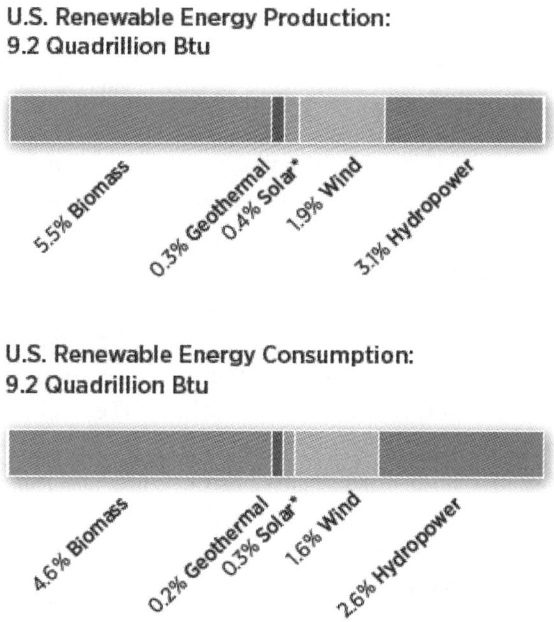

Figure 9.2. U.S. Renewable Energy Production and Consumption in 2013

****Solar PV (photovoltaic) data includes only on-grid systems of 1 megawatt or higher in capacity.***

Source: U.S. Department of Energy, 2014.

Figure 9.3. Top U.S. States for Renewable Electricity Capacity with Various Technologies in 2013

****Solar PV (photovoltaic) data includes only on-grid systems of 1 megawatt or higher in capacity.***

Source: U.S. Department of Energy, 2014.

Solar PV*
1 California
2 Arizona
3 New Jersey
4 North Carolina
5 Massachusetts

CSP
1 California
2 Arizona
3 Florida
4 Nevada
5 Colorado

Biomass
1 California
2 Florida
3 Georgia
4 Virginia
5 Alabama

Wind
1 Texas
2 California
3 Iowa
4 Illinois
5 Oregon

Geothermal
1 California
2 Nevada
3 Utah
4 Hawaii
5 Oregon

Hydropower
1 Washington
2 California
3 Oregon
4 New York
5 Alabama

According to data aggregated by the International Panel on Climate Change, life cycle global warming emissions associated with renewable energy, including manufacturing, installation, operation and maintenance, and dismantling and decommissioning, are minimal. Compared with natural gas, which emits between 0.6 and 2.0 pounds of carbon dioxide equivalent per kilowatt hour (CO_2E/kWh), and coal, which emits between 1.4 and 3.6 pounds of CO_2E/kWh, wind energy emits only 0.02 to 0.04 pounds of CO_2E/kWh, solar energy 0.07 to 0.2, geothermal energy 0.1 to 0.2, and hydroelectric power 0.1 to 0.5 CO_2E/kWh.[2] Renewable electricity generation from biomass can have a wide range of global warming emissions depending on the resource and how it is harvested. Sustainably sourced biomass has a low emissions footprint, while unsustainable sources of biomass can generate significant global warming emissions.

Use of renewable energy instead of fossil fuels offers significant public health benefits in terms of reduced air and water pollution, as emitted by coal and natural gas plants. Wind, solar, and hydro-source energy systems have no associated air pollution emissions. While geothermal and biomass energy systems emit some air pollutants, total air emissions are generally much lower than those of coal, petroleum, and natural gas sources. In addition, wind and some solar energy systems require essentially no water to operate and thus do not pollute water resources or strain supply by competing with agriculture, drinking water systems, or other important water needs. In contrast, fossil fuels can have a significant impact on water resources through the potential to pollute sources of drinking water and requiring large amounts of water usage in mining, extraction, processing, and energy conversions. Biomass and geothermal energy systems, like fossil fuel usage, do require water for operation. In addition, hydroelectric power plants impact river ecosystems both upstream and downstream from the dam. However, the National Renewable Energy Laboratory's "Renewable Electricity Futures Study" (2012), which included biomass and geothermal, found that water withdrawals and consumption would decrease by more than 50% by 2050 for these two technologies.[3]

9.2. Solar Energy

Solar energy utilizes the radiant heat and light from sun using a range of technologies that include solar heating, solar photovoltaic, and solar thermal power generation through concentrating solar power systems. Solar technologies are broadly characterized as either passive solar or active solar depending on the way they capture, convert, and distribute solar energy. Active solar techniques include the use of photovoltaic panels and solar thermal collectors to harness the energy; passive solar techniques include orienting a building to the sun, selecting materials with favorable thermal mass or light dispersing properties, and designing spaces that naturally circulate air.

Solar thermal energy (STE) is a form of energy and a technology for harnessing solar energy to generate thermal energy or electrical energy for use in industry and in the residential and commercial sectors. STEs can be used for water heating, space heating, space cooling, and process heat generation. A solar thermal collector is a device that collects heat from solar radiation; it commonly refers to solar hot water panels but may also refer to installations such as solar parabolic troughs and solar towers or basic installations such as solar air heaters. Concentrated solar thermal for process heat requirements in industries and concentrated solar power (CSP) for power generation utilize more complex collectors; simple evacuated tube collectors, glazed flat plate collectors, and unglazed plastic collectors are typically used in residential and commercial buildings for space and water heating.

As of 2013, the total global installed capacity of solar hot water systems was approximately 160 gigawatts (GW), with China as the world leader in their deployment with 70 GW installed and with Israel and Cyprus as the per capita leaders in the use of solar hot water systems with over 90% of homes using them. In the United States, Canada, and Australia heating swimming pools is the dominant application of solar hot water with an installed capacity of 18 GW. In the utility-scale, the 377-megawatt (MW) Ivanpah Solar Electric Generating System is the largest solar power plant in the world, located in California's Mojave Desert. The top countries with renewable electricity generation capacity with CSP technology in 2013 were Spain, the United States, the United Arab Emirates, India, and Algeria.

Photovoltaic (PV) systems generate electrical power by converting solar radiation into direct current electricity using semiconductors that exhibit the photovoltaic effect; the direct conversion of sunlight to electricity occurs without any moving parts or environmental emissions during operation. The technology is well proven, as photovoltaic systems have now been used for 50 years in specialized applications and grid-connected systems have been in use for over 20 years. Solar photovoltaic power generation has long been seen as a clean sustainable energy technology and, after hydro and wind power, is the third most important renewable energy source in terms of globally installed capacity. China, followed by Italy, Japan, and the United States, is now the fastest-growing market, while Germany remains the world's largest producer, contributing almost 6% to its national electricity demands. Solar electricity generating capacity grew by a factor of 35 between 2000 and 2013 in the United States and currently accounts for 0.5% of annual electricity generation.

9.3. Wind Energy

Wind energy converts the kinetic energy in wind into a useful form of energy, such as using wind turbines to produce electrical power, windmills for mechanical power, and wind pumps for water pumping or drainage. Wind power, as an alternative to fossil fuels, is plentiful, renewable, widely distributed, and clean, and it produces no greenhouse gas emissions during operation and uses little land. The effects on the environment are generally less problematic than those from other power sources. The major types of wind power are: utility-scale wind (wind turbines larger than 100 kilowatts are developed with electricity delivered to the power grid and distributed to the end user by electric utilities or power system operators); distributed, or "small," wind, which uses turbines of 100 kilowatts or smaller to directly power a home, farm, or small business as it primary use; and offshore wind, which are wind turbines erected in bodies of water around the world but not yet in the United States.

Worldwide there are now over 200,000 wind turbines operating in more than 80 countries, with a total operating capacity of around 320 GW as of 2013. The top countries with renewable electricity generation capacity with wind technology in 2013 were China, the United States, Germany, Spain, and India. In the United States, the wind power capacity is about 61 GW; the total number of operating utility-scale wind turbines at the end of 2013 was more than 46,000 and 40 states have operating utility-scale wind energy projects.

9.4. Hydro-Source Energy

Hydro-source energy is derived from the potential and kinetic energy of falling water and running water, which may be harnessed for useful purposes. Since ancient times, hydropower has been used for irrigation and the operation of various mechanical devices, such as watermills and sawmills; since the early 20th century, the term has been used almost exclusively in conjunction with the modern development of hydroelectric power with turbines at river dams. Hydropower has been an important source of flexible, low-cost, and emissions-friendly renewable energy. Used primarily to generate electricity today, the broad categories include: conventional dammed-hydro facility (hydroelectric dam); run-of-the-river hydroelectricity, which captures the kinetic energy in rivers or streams, without the use of dams; small hydro projects, which are 10 MW or less and often have no artificial reservoirs; micro hydro projects, which provide a few kilowatts to a few hundred kilowatts to isolated homes, villages, or small industries; conduit hydroelectricity projects, which utilize water that has already been diverted for use elsewhere, such as in a municipal water system; and pumped-storage hydroelectricity, which stores water pumped during periods of low demand to be released for generation when demand is high. The top countries with renewable electricity generation capacity with hydropower in 2013 were China, Brazil, the United States, Canada, and Russia. Hydropower remains the largest source of renewable electricity generation in the United States, primarily large-scale hydropower, which accounted for 6.6% of U.S. electricity generation in 2013.

9.5. Geothermal Energy

Geothermal energy is the heat from the earth that is drawn from several sources like hot water or steam reservoirs deep in the earth that are accessed by drilling, geothermal reservoirs located near the earth's surface, and the shallow ground near the surface, which maintains a relatively constant temperature of 10°–16°C (50°–60°F); this heat can be utilized as an energy source on both large- and small- scales. A utility can use the hot water and steam from reservoirs to drive generators and produce electricity. Other applications apply the heat produced from geothermal directly to various uses in buildings, roads, agriculture, and industrial plants (direct use). Still others use the heat directly from the ground to provide heating and cooling in homes and other buildings (heat pumps). Geothermal power is cost effective, reliable, sustainable, and environmentally friendly, but it has historically been limited to areas near tectonic plate boundaries with seismic activity. Recent technological advances have dramatically expanded the range and size of viable resources, especially for applications such as home heating and cooling. Geothermal wells release greenhouse gases trapped deep within the earth, but these emissions are much lower per energy unit than those of fossil fuels. As a result, geothermal power has the potential to help mitigate global warming if widely deployed in place of fossil fuels.

The International Geothermal Association (IGA) has reported that about 11,000 MW of geothermal power in 24 countries is online as of 2013, representing a 20% increase in capacity since 2005, and projects growth to 18,500 MW by 2015. The United States leads the world in installed geothermal electricity capacity and generation, with most of that capacity installed in California, followed by Nevada. In 2013, the U.S. geothermal electricity production capacity was about 3,800 MW from 77 power plants. The Philippines is the second-highest producer, with 1,900 MW of capacity online, followed by Indonesia, Mexico, and Italy.

9.6. Biomass and Biofuels

Biomass, which is biological material derived from living or recently living organisms and most often refers to plants or plant-based lignocellulosic materials, can be used either directly via combustion to produce heat or indirectly after converting it to various forms of biofuels. Conversion of biomass to biofuel can be achieved by different methods that are broadly classified into thermal, chemical, and biochemical methods. Wood remains the largest biomass energy source to date; it includes forest residues (such as dead trees, branches, and tree stumps), yard clippings, wood chips, and even municipal solid waste. Industrial biomass can be grown from numerous types of plant species, including switchgrass, poplar, willow, sugarcane, and bamboo. For grain crops, the grain can be used for liquid transportation fuels, while the straw can be burned to produce heat or electricity. Plant biomass can also be degraded from cellulose to glucose through a series of chemical treatments, and the resulting sugar can then be converted to biofuel. Additionally, biomass can be converted to other usable forms of energy, such as methane gas (landfill gas [LFG] or biogas) or such transportation fuels as ethanol and biodiesel. Crops, such as corn and sugarcane, can be fermented to produce the transportation fuel ethanol. Biodiesel, another transportation fuel, can be produced from leftover food products like vegetable oils and animal fats. Technologies such as biomass to liquids, cellulosic ethanol, and algae-derived biomass and biofuels are under active research and development.

Biofuels are often categorized into conventional biofuels, which utilize well-established technologies and are in commercial production today, and advanced biofuels, which are under development or in pilot and demonstration phases. Advanced biofuel technologies could have the potential benefits of high land-use efficiency and use of nonarable land for their production.

Conventional biofuels:

- Ethanol from corn, sugarcane, wheat, sugar beet, cassava
- Biodiesel from soybean, canola (rapeseed), oil palm, sunflower
- Biogas produced via anaerobic digestion of energy crops (e.g., maize silage) and wastes (e.g., organic waste, manure)

Advanced biofuels:

- Cellulosic-ethanol from different types of lignocellulosic biomass
- Biomass-to-liquids like diesel from different types of lignocellulosic biomass
- Biosynthetic gas like biomethane produced from different types of bio mass via gasification
- Other novel technologies in the R&D phase, such as algae-biofuels and sugar-based hydrocarbons

Biomass can be used for fuels, power production, and products that would otherwise be made from fossil fuels. In such scenarios, biomass and biofuels can provide an array of benefits. The use of biomass energy has the potential to greatly reduce greenhouse gas (GHG) emissions. Burning biomass releases about the same amount of carbon dioxide as burning fossil fuels; however, while fossil fuels release essentially "new" greenhouse gas, biomass releases carbon dioxide that is largely balanced by the carbon dioxide captured in its own growth (depending on how much energy was used to grow, harvest, and process the fuel). While biomass and biofuels can reduce CO_2 emissions, compared with the use of fossil fuels like coal and petroleum, uncertainty on the impact of land-use change, food security, and environmental sustainability (deforestation, indirect land-use change emissions, biodiversity impact, and water consumption) remains an important topic that continues to be questioned and needs to be addressed.

In 2013, biomass electricity generation accounted for 11% of all renewable energy generated in the United States. Biomass electricity came primarily from wood and agricultural residues that were burned as a fuel for cogeneration (simultaneous production of electricity and heat) in the industrial sector (such as in the pulp and paper industry). Other top countries in renewable electricity generation from biomass are Germany, China, Brazil, and India.

In relation to biofuels used for transportation, ethanol and biodiesel are the most established commercially in the United States. In 2013, the United States produced 57% of the world's ethanol, followed by Brazil at 27%, the European Union at 6%, China at 3%, and Canada at 2%. U.S. corn ethanol production, in contrast to Brazil's ethanol production from sugarcane, has expanded rapidly between 2000 and 2013 by a factor of eight; U.S. ethanol production totaled about 50.3 billion liters (13.3 billion gal) per year in 2013, with Iowa being the largest producer. In 2013 the United States led the world in biodiesel production, followed by Germany, Brazil, Argentina, France, and Indonesia. U.S. biodiesel production expanded to 6.8 billion liters (1.8 billion gal) in 2013, a 63% increase from 2012 and 360 times what it was in 2001.

9.7. Waste-to-Energy

Energy recovery from waste is the conversion of nonrecyclable waste materials into usable heat, electricity, or fuel through a variety of processes, including combustion, gasification, pyrolization, anaerobic digestion, and LFG recovery.[4] This process is often called waste-to-energy (WTE) and is a form of energy recovery. Converting nonrecyclable waste materials into electricity and heat generates a renewable energy source and reduces carbon emissions by offsetting the need for energy from fossil sources and reduces methane generation from landfills. The common WTE processes today produce electricity and/or heat directly through combustion or produce a combustible fuel commodity, such as methane, methanol, ethanol, or synthetic fuels.

9.7.1. Incineration

Incineration, the combustion of waste with energy recovery, is the most common WTE implementation. All modern WTE plants incinerating waste must meet strict air emission and residual ash management standards. The upfront capital needed to build a municipal solid waste (MSW) incineration plant can be a significant hurdle when building a new facility, and the economic benefits of MSW incineration may take several years to be fully realized. Long-term contracts (generally 30 years) are often developed between the facility and the municipality to secure a guaranteed waste stream. MSW incineration facilities typically collect a tipping fee from the independent contractors that drop the waste off on a daily basis. The facilities also receive income from utilities after the electricity generated from the waste is sold to the grid. A possible third stream of revenue for the facilities comes from the sale of both ferrous (iron) and nonferrous scrap metal collected from the post-combusted ash stream.

Currently there are 86 facilities located in 25 states in the U.S. for combustion of municipal solid waste, with energy recovery. These facilities have the capacity to produce 2,720 MW of power per year by processing more than 25 million metric tons (28 million tons) of waste per year. There are three main types of technologies for the combustion of municipal solid waste:

- **Mass burn facilities:** Mass burn facilities are by far the most common types of combustion facilities in the United States. The waste used to fuel the mass burn facility may or may not be sorted before it enters the combustion chamber. Many advanced municipalities separate the waste on the front end to pull off as many recyclable products as possible.
- **Modular systems:** Modular systems are designed to burn unprocessed, mixed MSW. They differ from mass burn facilities in that they are much smaller and are portable. They can be moved from site to site.
- **Refuse-derived fuel systems:** These facilities use mechanical methods to shred incoming MSW, separate out non-combustible materials, and produce a combustible mixture suitable as a fuel in a dedicated furnace or as a supplemental fuel in a conventional boiler system.

9.7.1. Emerging WTE Technologies

There are a number of new and emerging technologies that are able to produce energy from waste and other fuels without direct combustion. Many of these technologies have the potential to produce more electric power from the same amount of fuel than would be possible by direct combustion. Some are able to efficiently convert the energy into liquid or gaseous fuels:

Thermal technologies:

- Gasification (produces combustible gas, hydrogen, synthetic fuels)
- Thermal depolymerization (produces synthetic crude oil, which can be further refined)

- Pyrolysis (produces combustible tar/bio-oil and chars)
- Plasma arc gasification (produces rich syngas [synthesis gas] including hydrogen and carbon monoxide usable for fuel cells or generating electricity)

Nonthermal technologies:

- Anaerobic digestion (Biogas rich in methane)
- Fermentation production (examples are ethanol, lactic acid, hydrogen)
- Mechanical biological treatment (MBT)

9.8. Industrial Applications of Sustainable Energy

In recent years, renewable energy has increasingly attracted public and policy attention particularly for its potential to contribute to reductions in GHG emissions. Most interest has focused on the use of renewables in power generation and as biofuels. Although some attention has been paid to the potential for renewables, particularly biomass and solar thermal technologies, to contribute to heating and cooling in residential and commercial space heating applications, their use in industrial applications has received less interest. Renewable energy plays a relatively small role in industry today; about 7% of U.S. total industrial energy consumption in 2013 was from renewables consisting primarily of wood and wood-derived fuels, municipal solid waste, and conventional hydropower.[1] According to the International Renewable Energy Agency (IRENA), "More than 80% of the total final consumption . . . in the manufacturing industry is estimated to be used as process energy by 2030. . . . The energy efficiency improvement potential alone are [*sic*] not sufficient to reduce the increasing demand of fossil fuels and their associated environmental impact."[5] Achieving higher penetration levels for renewable energy in industry is key to realizing substantial reductions in the sector's fossil fuel demand and related CO_2 emissions, and more efforts are required to examine the role of renewable energy in improving sustainability of the manufacturing industry sector. Renewable energy technologies can provide practical and cost-effective alternatives for process heat generation and as a carbon source for the production of chemical and plastics.

Today mainly the pulp and paper industry and the food processing industry use renewable energy resources, accounting for approximately 45%

and 25% of their energy demand, respectively. Biomass currently makes by far the most significant renewable energy contribution to industry, providing around 8% of its final energy use; the use of biomass, primarily for process heat, has been most prevalent in the pulp and paper and the wood product sectors.

Some of the barriers to adoption of sustainable energy usage in the industry, as identified by the United Nations Industrial Development Organization (UNIDO), include:[6]

- Lack of information on the potential contribution of renewables and ways of achieving it
- Cheap fossil fuels
- Absence of appropriate technology supply chains
- Lack of technical capacity
- High cost of capital in many developing countries
- Focus on upfront investment cost instead of full life cycle cost
- Risks associated with technology transitions and the adoption of early stage technologies
- Restricted access to financial support to cover the extra costs of best available technology (BAT)
- The lock-in of inefficient, polluting technologies with long lifetimes

The successful deployment of climate-friendly technologies depends heavily on firms having the financial capacity to invest in the relevant technologies. But it also depends on their having the knowledge to develop and exploit such technology and on the existence of organizational structures and cultures, including institutional settings and rules that encourage the use of such technology. Site-specific issues will also influence the most appropriate climate-friendly technologies for a particular industrial application. To maximize efficiencies, the demand for process energy inputs must be minimized through the introduction of low-cost, efficient technologies and through systems optimization before renewable energy options are applied.

Based on the findings of the techno-economic analysis and the feedback from stakeholders, IRENA identified six priority areas that warrant action from both policy makers and industrial stakeholders. These are:[5]

- **Energy-intensive sectors:** With 75% of the total industrial energy demand and long lifetimes for these types of plants, the energy-intensive sectors, namely iron and steel, chemical and petrochemical, nonferrous metals, nonmetallic minerals, and pulp and paper, need to consider renewable energy options not only as an integral part of their new build capacity but also as part of their existing capacity.
- **Small- and medium-size enterprises (SMEs):** Accounting for more than 90% of all manufacturing businesses, SMEs play a crucial role in increasing the deployment rate of renewable energy technologies, providing local manufacturing opportunities, and stimulating cost reductions through learning by doing.
- **Biomass:** Among the renewable technology options, biomass has the largest substitution potential in the manufacturing industry, but immediate and internationally coordinated action is required to alleviate the serious supply constraint of sustainably sourced and low-cost biomass resources and to deploy the most resource-efficient biomass use applications.
- **Solar thermal systems:** Solar thermal heat systems have a large technical and economic realizable potential in small-scale plants and less energy-intensive industries such as the textile and food sectors, but the vicious circle of high initial capital costs and low deployment rates needs to be broken.
- **Electrification:** With increased electrification in the industry sector, renewable energy deployment can be achieved only through technology development in both the industry and power sectors.
- **Regional aspects:** Regional potential depends on production growth, ratio of existing and new capacity, and renewable resource availability. Energy pricing and climate policies can ensure a level playing field and biomass resource constraints may be elevated by trade, but equally important will be specific policies to support the different industries in deploying renewable energy.

Action Areas for Energy-Intensive Industries

- **Renewable energy integration into new capacity and maximization of the potential in existing capacity:** The long life of plants (up to 60 years and even more) and large energy demands are main challenges for renewable energy deployment. This requires continuous access to energy supplies (e.g., biomass) and in most cases on-site land for storage of biomass or construction of solar thermal plants. Another barrier is that most production processes in energy-intensive plants (e.g., chemical plants) are optimized with highly integrated material and energy flows. When designing new plants, renewable energy technologies need to be exploited. Existing plants where renewable energy technologies can be integrated without major process modifications need to be identified.

- **Renewable energy technologies for high-temperature heat demand:** High-temperature heat applications account for more than two-thirds of the total process heat demand. Currently, biomass offers the only renewable energy option to provide high-temperature heat. To an extent, solar thermal heat systems could also provide high-temperature heat, but more research, development, and demonstration (RD&D) is required to commercialize those technologies available.
- **Sustainable biomass logistics:** Most production facilities are large-scale centralized plants operating at economies of scale, which will require large energy flows to be brought from within and across national borders. The key factors defining the economic viability of biomass are its energy density, its production costs, the distance it needs to be transported, and the type of transport mode. Cost competitiveness of biomass can be maintained through an effective logistics infrastructure as consumption increases.

Action Areas for SMEs

- **Overcoming barriers to new technology uptake:** Energy costs often form a substantial part of the overall production costs, yet few SMEs know how they can reduce their dependence on expensive fossil fuels. Additionally, most production capacity is based on aging and inefficient equipment, thereby increasing the energy demand of process heat generation. Replacing inefficient equipment with new capacity and integrating new capacity with renewable energy technologies are two key areas that can benefit SMEs. To realize the many opportunities that exist for the SMEs, barriers will need to be eliminated by developing effective strategies.
- **SME clusters:** Many SMEs operate in clusters organized around a specific set of industrial operations. The proximity of multiple SMEs with similar operations and energy demand structures makes SME clusters a target to promote renewable energy deployment.
- **Mix of renewable energy technologies:** The low absolute energy demand per plant in most SMEs and the varying temperature levels of process heat

use offer the opportunity to take advantage of the range of renewable energy technologies available. Biomass, solar, geothermal, and heat pumps can all be deployed in SMEs.

Action Areas for Biomass

- **The sustainability of biomass resources:** Biomass is at the heart of the nexus of energy, food, water, and land scarcity, offering the highest potential for deployment in all three end-use sectors of transport, buildings, and industry. To reach this potential by creating the least additional burden on water and land sources, biomass needs to be sourced in a sustainable manner. One sustainable source of biomass is biogenic waste material; however, inconsistent waste regulation, cheap waste disposal options, and the lack of an infrastructure to collect, transport, and use waste flows limit the use of waste as a resource. Residues can provide an important share of the potential demand, although their availability may not be sufficient. Energy crops could provide the remaining demand, but they need to be cultivated sustainably.

- **Trade barriers for biomass:** Existing producer and consumer markets for manufacturing products are not necessarily found in biomass-rich regions. Consequently in some territories, trade barriers and taxes are important obstructions to the deployment of renewable-based products. These should be regulated in a way that creates opportunities for the deployment of renewable energy technologies.

- **Cascading use of biomass:** Cascading the use of biomass will allow its use as a feedstock to bio-based polymers first, after which the discarded products can be recycled for several life cycles and subsequently incinerated with heat and electricity coproduction at the end of their life. Compared with immediate burning of biomass for heat or power generation, cascading offers the possibility to use the same resource for multiple purposes. Making the best use of limited biomass resources requires recycling rates to be increased by reducing the quantities of plastic ending up in landfills, maximizing recycling trips of plastics to the extent that the material properties allow, and incinerating postconsumer waste with efficient energy recovery. This would reduce the competition for biomass resources, create added economic value, and allow for multiple uses of the same resource.

Action Areas for Solar Thermal Process Heating

- **Expansion of deployment:** Solar thermal systems have been successfully distributed within the building sectors and resulted in greater local manufacturing capacities and reduced costs. Similar deployment strategies should be used throughout industry, starting with targeting low- and medium-temperature heat applications. Additionally, the deployment of advanced vacuum tube collectors with compound parabolic concentrator or the recently developed high vacuum flat-plate collectors for high-temperature applications should be promoted.
- **Focus on SMEs:** SMEs are an interesting market for solar thermal systems for three reasons. First, SMEs often rely on volatile and expensive fossil fuels for heat production, and solar thermal systems reduce their dependency on fossil fuels and contribute to the reduction of operating compared with the energy-intensive industry, which makes the integration of solar heating systems easier. Third, the sheer number of SMEs present could result in rapidly declining costs, owing to learning by doing and more effective operations by installers. This could create a virtuous circle in which with declining costs and more experience, the deployment of solar heating systems is accelerated in this market segment.
- **Local manufacturing:** The current costs of industrial solar thermal systems are determined by a relatively small number of suppliers of these highly sophisticated technologies. Although these technologies have a high performance, they are generally too costly for a global market. More equipment being manufactured locally will reduce capital costs and create added value and local business opportunities along the supply chain.

Action Areas for Electrification

- **Decarbonization of the power sector:** Increased electricity consumption from the industrial sector should, wherever possible, be provided by the development of new renewable power generation capacity. Renewable power generation costs have fallen for wind technology, solar photovoltaic (PV), concentrated solar power, and some biomass technologies, although hydropower and geothermal produced at good sites remain the cheapest ways to generate electricity.

- **Relocation:** New capacity investments in electricity-intensive industry sectors should, wherever possible, be located near renewable energy power generation sites. This will reduce the transmission and distribution losses associated with electricity transmission and will simultaneously provide a reliable and affordable energy source for these plants.
- **On-site renewable energy electricity generation:** Several large manu facturing companies are integrating renewable energy power generation into their existing manufacturing plants either through solar PV panels on the production facilities, or wind turbines on site, or other sources of renewable energy.
- **Technology development:** Process technology R&D should also focus on electricity-based alternatives, ensuring that the electricity sector is decarbonized.

Action Areas for Regional Aspects

- **Create a level playing field for renewable energy alternatives:** Profitability and costs will remain the main drivers for both consumers and companies. The inclusion of externalities, such as CO_2 emissions (in particular, given the emissions from the carbon-intensive coal used by many industries), in the price of products at a global level will make renewable-based products more competitive. The introduction of a CO_2 price or other forms of pricing could be considered in order to support the deployment of renewable-based products. In addition, low cost coal is a major potential barrier to the limited deployment of renewable energy. Removing fossil fuel subsidies and ensuring the availability of low cost renewable energy technology options can help support a transition to a less fossil fuel–dependent manufacturing industry such as exists in China and India.
- **Biomass trade:** Several countries such as China have a limited biomass resource relative to their energy consumption, while regions like Latin America and Africa have abundant biomass resources. Fortunately, neighboring countries of China have rich sources of biomass (e.g., India and other developing Asian countries). Importing excess biomass from these regions would increase the supply in China and overcome the barriers related to the limited domestic production. Biomass trade networks could also be developed to export the unutilized biomass from Africa and Latin America to regions that need biomass for their industrial sectors.

- **Make abundant renewable energy resources available for industry:** Reliable and affordable energy is an important factor for manufacturing industries when choosing their next production facility. Through the development of its abundant renewable energy resources, Africa has the ability to attract new industries and build a truly green economy.
- **Use existing resource flows:** Many of Africa's industrial sites use biomass-based resources as input into their production processes. Examples are the food processing industry (tea, coffee, cocoa and dairy), the sugar industry, the textile and yarn industry, and the saw and wood processing industry. Industrial waste streams from these industries can be effectively used to meet the energy demand of the very same industries.
- **Accelerate technology development for modular production:** In the mid- to long-term, there are opportunities to expand the renewable-based production in those regions that are resource rich, such as Asia and Africa. This will require technologies that are able to handle the local mix of renewables being available, as well as new technologies for the currently underexploited biomass resources. Widespread expansion into Asia and Africa would also require technologies that operate in a more modular approach, whereby production takes place on the basis of locally available resources.

References:

1. "2013 Renewable Energy Data Book," National Renewable Energy Laboratory, U.S. Department of Energy, www.nrel.gov/docs/fy15osti/62580.pdf, retrieved December 2014.
2. "Special Report on Renewable Energy Sources and Climate Change Mitigation," Intergovernmental Panel on Climate Change, http://srren.ipcc-wg3.de/report/, retrieved September 2014.
3. "Renewable Electricity Futures Study," National Renewable Energy Laboratory, http://www.nrel.gov/analysis/re_futures/, retrieved September 2014.
4. "Energy Recovery from Waste," U.S. Environmental Protection Agency, www.epa.gov/waste/nonhaz/municipal/wte/, retrieved December 2014.
5. "Renewable Energy in Manufacturing," International Renewable Energy Agency (IRENA), http://irena.org/remap/REmap%202030%20Renewable-Energy-in-Manufacturing.pdf, retrieved December 2014.
6. "Renewable Energy in Industrial Applications—an Assessment of the 2050 Potential," United Nations Industrial Development Organization (UNIDO), www.unido.org/fileadmin/user_media/Services/Energy_and_Climate_Change/Energy_Efficiency/Renewables_%20Industrial_%20Applications.pdf, retrieved September 2014 and June 2015.

Chapter 10

Green Supply Chain, Logistics, and Transportation

Consumer awareness and sustainability legislation have pushed environmental issues into the spotlight, making it vital for manufacturing systems to have a plan of action for "going green." The different environmental conscience of industrial businesses and manufacturing systems can be positioned along their environmental maturity level. Manufacturing systems can be placed in one of three environmental management maturity levels, as follows:

1. Reactive stage: Manufacturing systems at this phase are worried about the cost of failure to act in accordance with environmental legislation; hence, the emphasis is on the end of the process, for instance, the use of filters in chimneys or the correct disposal of waste. So, environmental management does not occur within the production process with the intent of performing some type of structural change.
2. Preventive stage: Manufacturing systems at this phase prevent/minimize the generation of waste or the overuse of inputs. Here the emphasis is on the pursuit of eco-efficiency.
3. Proactive stage: Manufacturing systems at this phase take various features of environmental management into consideration as strategies for gaining competitive advantage. The objective of environmental management is to make substantial modifications in the processes and products. Proactive companies are reaping benefits in the form of cost savings, favorable public opinion, and accessibility to clean energy stimulus funds.

Environmental system management cannot be restricted to the manufacturing systems and organizations, as the whole system influences and is affected by the supply chain. Manufacturing systems have various interdependencies, since they are involved in material and information flows from the supplier of raw materials to the final customer. This broader supply chain standpoint is significant in the transition toward sustainability, and it is termed green supply chain management.

Logistics and transportation are the integrated management of all the activities required to move products through the supply chain. The logistical activities entail freight transport, storage, inventory management, materials handling, and all the associated information processing. The main objective of logistics is to coordinate these actions in a way that meets customer requirements at minimum cost. In the past, this cost has been expressed in merely financial terms. As concern for the environment rises, manufacturing systems should take more account of the external costs of logistics related chiefly to generating chemical, liquid, and solid wastes; climate change; air pollution; noise; vibration; accidents; and so forth.

Integrating environmental sustainability performance into logistics management is still a new phenomenon. Green logistics and transportation attempts to move and deliver raw materials and products at the lowest possible cost while maintaining the highest standards and minimizing environmental impact in the process. It implies innovation in all steps of the supply chain, the conception of products and in some cases the final use of products. Green logistics management reveals the capability to conserve resources, reduce waste, enhance operational efficiency, and meet the social expectation of environmental protection. Green logistics is designed to be not only environmentally friendly but also economically functional.

10.1. Green Supply Chain Network and Green Supply Chain Management Concepts

The green supply chain (GSC) network concept has been coming into sight since 2010. Adjoining the "green" concept to the "supply chain" concept generates a new paradigm where the supply chain will have a direct relation to the environment. *GSC network* is defined as the incorporation of environmental thinking into the management of the supply chain, containing product design, procurement, material selection, and product delivery to the end customer as well as controlling the life cycle of the product's afterlife. *GSC management* can be defined as the coordination of the supply chain in a form that integrates environmental concerns and considers interorganizational activities. GSC management copes with the acquisition, production, and distribution of materials to meet the requirements of stakeholders to enhance profitability, competitiveness, and the resilience of the suppliers, manufacturing systems, distribution centers, and retailers in the short and medium terms through improved green performance.[1] Improvements by GSC management can be categorized as follows:[2]

- Operations improvement: GSC management improves the operations by incorporating environmental and waste managerial solutions.
- Agility enhancement: GSC management helps mitigate risks and speed innovations.
- Adaptability improvement: GSC analysis often leads to innovative processes and continuous improvements.
- Alignment improvement: GSC management involves negotiating policies with suppliers and customers, which results in better alignment of manu facturing processes and principles.

To be more sustainable, a rising number of producers and manufacturers are looking further into the supply chain, at the suppliers of the parts, and at the equipment used on a daily basis in their sites. By working with partners that understand the meaning of sustainability, manufacturers can further enhance their credentials at every point of the supply chain. Since the concept of GSC management was first put forward in the Manufacture Research Consortium, at Michigan State University in 1996, domestic and foreign academe embarked on studying different aspects of GSC management[3]

Recognizing that business would implement GSC practices only if it were genuinely to expect to obtain some significant benefits, Bowen et al., as pioneers in performing research in the area of GSC management, conducted a two-phase survey of GSC activities in the United Kingdom in 2001 for a sample of business units. In their research, they identified three types of GSC initiatives involving businesses that collect environmental information on suppliers and make assessments of the performance of those suppliers, businesses that make changes to the end product supplied, and businesses that incorporate joint clean technology initiatives with suppliers. They requested that purchasing managers answer some questions related to the identification of implementing any of the green supply initiatives in the interviews. For each initiative, managers were asked whether they had

Supplier Environmental Management

Supplier:

Site:

Responsible person:

Function:

Telephone:

Telex:

1. Does your company have an environmental management system in place in accordance with:

☐ yes ☐ no Remarks:

☐ Environmental management and audit scheme or environmental working group 1836/93
☐ ISO 14001
☐ Internal environmental management system

If yes, please provide a copy of your certificate.

Not yet certified, but planned for in 20--.

☐ Environmental management and audit scheme or environmental working group 1836/93
☐ ISO 14001
☐ Internal environmental management system

2. Does your company have another management system that includes environmental management?

☐ yes ☐ no Remarks:

If yes, what kind?

3. Are environmental protection measures audited in your company?

☐ yes ☐ no Remarks:

If yes, by whom?

☐ Internal Auditor/s ☐ External Auditor/s

4. In your company what kinds of processes are there?

☐ Production processes regarding parts/components
☐ Production processes regarding final products
☐ Assembling processes
☐ Disposal and recycling processes

Remarks:

5. Are the processes regularly inspected and assessed?

☐ yes ☐ no Remarks:

Please select the related processes that are inspected:

☐ Production processes regarding parts/components
☐ Production processes regarding final products
☐ Assembling processes
☐ Disposal and recycling processes

Remarks:

6. Are environmental aspects an integrated factor in your product and process planning?

☐ yes ☐ no Remarks:

7. Does your company strictly adhere to relevant environmental regulations?

☐ yes ☐ no Remarks:

8. Are environmental measures and the results documented in your company?

☐ yes ☐ no Remarks:

9. Has your company established goals for continual improvement of performance, and is the achievement of these goals documented?

☐ yes ☐ no Remarks:

10. Are your company's employees informed and trained on environmental issues?

☐ yes ☐ no Remarks:

11. Are you working with your suppliers of raw materials/parts and contractors to improve their environmental performance?

☐ yes ☐ no Remarks:

Please briefly describe the environmental impact associated with your companies' processes

Signature Date

Figure 10.1. Overview of a Qualitative Survey for Assessing the State of Environmental Management in Suppliers' Operations

been "implemented," "planned," or "not planned" in that operating unit. Finally, they concluded that the most prevailing supplier-related initiative is the product-based strategy, where the company wants to minimize waste, excess cost, and resources from the sourcing process at the customer-supplier interface.[4] Over the past decade, car manufacturers have stepped up their efforts to green their operations and their supply chains, partly in response to increasingly stringent environmental legislation and partly in response to increasing customer expectations concerning the minimization of fuel consumption, the reduction of harmful emissions, and the increase in recycling. As a case study, the BMW company has influenced the environmental management of suppliers by pushing forward the eco-design of its products. It is essential for BMW to find suppliers that proactively assist in designing sustainable products. For example, if fuel consumption has to be reduced, the vehicle design must be more lightweight, which requires those that supply the components to source and develop appropriate materials. In many cases, automotive manufacturers such as BMW now include an environmental dimension in their vendor rating systems, systems that previously would have focused solely on performance measures such as quality and cost. BMW works jointly with suppliers to optimize not only quality and cost but also environmental performance. An overview of a qualitative survey for assessing the state of environmental management in the suppliers' operation is shown in Figure 10.1.

Currently, while there is a lot of interest in the concept of GSCs, a gap still exists between the theory and the implementation of the concept in the industry. Greening the supply chain helps companies to avoid potential environmental problems that might arise with their suppliers and that in turn could threaten their own environmental performance. Strategic environmental sourcing provides firms and manufacturing systems with opportunities to design products that are more environmentally friendly while at the same time reducing waste, risk and costs. In the past, purchasing was considered a clerical function related to manufacturing and materials management. Slowly, it became a tactical function of marketing and manufacturing, and now it is considered part of corporate strategy, helping the firm to be involved in the environmental opportunities found in supply chain management. Supplier quality assurance and assessment programs, supplier development programs, and supplier measurement programs are some parts of this strategy. For many firms and manufacturing systems getting started in the assessment of a supplier's environmental performance, new metrics will be necessary. Examples of environmental performance indicators that can be considered as a good starting point are as follows:

- The percentage of biodegradable/compostable material, the amount of emissions and waste (per unit of product), the percentage of volatile organic compound (VOC) content, pre-/post-consumer recyclable content (%), the amount of solid waste and total energy used, having access to the secondary markets and recycling centers for recyclable and reusable waste generated
- No ozone-depleting substances, no hazardous air emissions, no hazardous waste, no toxic pollution, use of less hazardous alternative (% of weight/volume)
- Commitment to periodical environmental auditing, having ISO 14000 certification, having an energy efficiency label related to the procured materials or products, environmentally responsible packaging, take-back or reverse logistics program, public disclosure of environmental record
- Number of hours related to training sessions for teaching environmental practices per employee, participation in voluntary Environmental Protection Agency (EPA) programs

Along with the GSC management concept, the 10 most important indicators of supplier environmental performance can be categorized as follows:

1. Public disclosure of environmental records, such as corporate environmental reports and toxic release inventory data.
2. ISO 14000-certified: having third-party certification of environmental management systems.
3. Second-tier supplier environmental evaluation: evaluation of suppliers beyond the first-tier (direct contact) firms.
4. Reverse logistics program: a system in place for the recovery of products or packaging from the consumer or supply chain members.
5. Hazardous waste management: a process in place to document and manage hazardous materials and waste.
6. Environmentally friendly product packaging: use of recycled materials for packaging or waste minimization of packaging materials.

7. Toxic pollution management: a process in place to document and manage toxic pollution.

8. Controlling ozone- depleting substances, substances that when used contribute to the degradation of ozone.

9. On EPA 17 Hazardous Material List: does the firm use any of the chemicals on the EPA hazardous materials listing?

10. Hazardous air emissions management: a process in place to document and manage air emissions.

In the next section, we will focus on definitions of closed-loop supply chains and closed-loop manufacturing resource planning.

10.2. Closed-Loop Supply Chains

Based on the literature review, and recently owing to growing attention given to the implementation of sustainability in manufacturing systems in terms of reduction in the primary resource consumption, pollution prevention, waste management, social responsibility, and the customer pressures, reverse logistics in supply chain management has gained lots of consideration. Since planning and designing the forward and reverse logistics separately results in suboptimal designs with regard to the objectives integrated with the supply chain, the closed-loop supply chain (CLSC) network designing and planning are analytically essential. CLSC can assure the least waste of materials through the whole supply chain during the life cycle of the materials.[5, 6]

A simple structure of a proposed CLSC network is demonstrated in Figure 10.2. In the forward stream, the suppliers provide the raw material to manufacturing facilities. The products are transported from plants to customers via distribution centers to fulfill the customers' demands. In the backward stream, the returned products from customers are shipped to collection centers for testing and inspection.

After examining the products in collection centers, the reusable parts of the products are transported to the sites, the repairable products are shipped to the repair centers, and the recyclable products and wastes are sent to recycling centers and disposable products are dispatched to the disposal centers. In this way, redundant transportation of the returned products is reduced and the returned products are directly transferred to the relevant facilities. After fixing the products in repair centers and remanufacturing

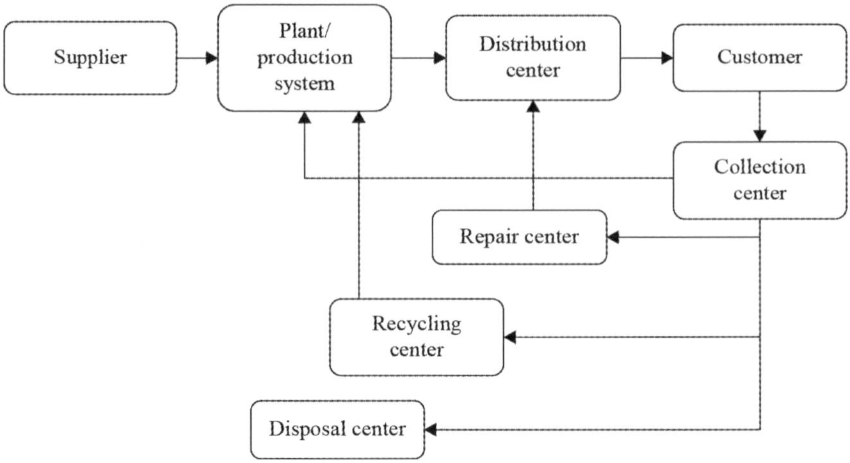

Figure 10.2. Simple Structure of a Proposed Closed-Loop Supply Chain Network

in the plants, the returned products are transported to distribution centers. When the repair and remanufacturing activities are made, the products are of as-good-as-new quality; hence, they can be used in various related industries. The proposed model can take the following assumptions and constraints into account:

- The model can be single-product or multi-product.
- The model can be single-period or multi-period. (Here, the term *period* refers to a given specific time or a timing cycle that is needed to perform all of the system's processes, containing production, recycling, and repairing of products; controlling the inventory; and meeting all or some parts of the customers' demands.)
- The locations of suppliers, customer zones, and disposal centers can be known and fixed. It is also possible to have various alternatives for disposal centers' locations and attempt to select the best location based on initial investments, closeness to the markets and recycling centers, and so forth.
- The potential locations of plants, distribution centers, collection centers, and repair centers are known.
- The capacities of facilities are limited.
- The quantities of all parameters can be deterministic or probabilistic.

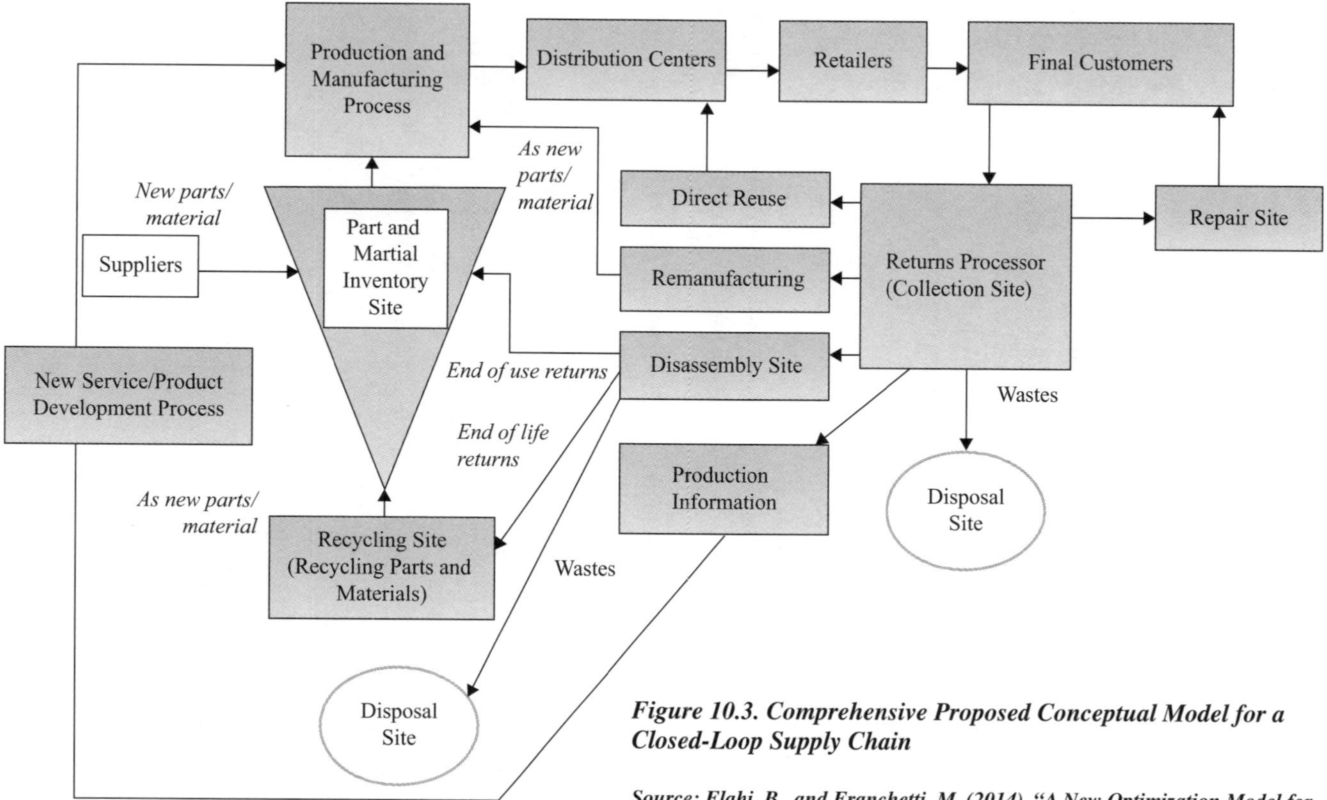

Figure 10.3. Comprehensive Proposed Conceptual Model for a Closed-Loop Supply Chain

Source: Elahi, B., and Franchetti, M. (2014), "A New Optimization Model for Closed-Loop Supply Chain Networks."

- The return rate for each customer can depend on the customer demand.
- The objective can be defined as determining the facility locations and flows between facilities along with financial decisions.

Elahi and Franchetti proposed a comprehensive conceptual model for a closed-loop supply chain in 2014, as is illustrated in Figure 10.3.[6] In this conceptual model, product life cycle and three types of returns are taken into consideration as well. Once products are applied by final customers, some of them are returned back. The returning products are delivered to the collection site. Commercial returns are fixed at the repair site. These products can be utilized as new ones. End-of-use and end-of-life returns are disassembled. In this stage, the wastes are separated. End-of-life returns are recycled at recycling sites. The parts are added to the part inventory as new parts.

It should be considered that the capacities of manufacturing, repairing, disassembling, and recycling sites/centers can be limited. Based on the number of returning parts, the manufacturer purchases new parts from external suppliers. There are some suppliers who can supply required parts. The capacities of suppliers can be known. Some of the returned parts are not usable and should be disposed of. The commercial returns can supply a portion of market demand.

10.2.1. Closed-Loop Manufacturing Resource Planning

Closed-loop manufacturing resource planning (CLMRP), also recognized as closed-loop resource planning, refers to a manufacturing resource planning model that includes returned or recycled products or components as part of the supply chain. The conception of CLMRP assures a broad range of paybacks and benefits, such as:

- Cost savings associated with greener and cleaner facilities, operations, and production systems
- Benefits integrated with regulations compliance to lessen and level the raw materials shortages
- Savings on virgin material
- Revenue on recovered material
- Cost differential between new and used parts
- Savings on landfill tipping fees

- Long-term relationships with customers and suppliers in order to enhance loyalty of customers and increasing paybacks; decrease lead times and expedite the replenishment process related to procuring material, parts, and products by suppliers; and focus on obtaining more benefits for all of the involved members (echelons) via efficient collaboration and cooperation through the entire system
- Environmental benefits such as having green image, reduction of solid/liquid wastes, climate change mitigation

CLMRP is also related to any proactive end-of-life (EoL) strategy that keeps a manufacturer's product out of a landfill or from being incinerated. The hierarchy of EoL activities can be ordered as follows:

1. Prevent discarding
2. Reuse of the products
    a) Reuse as complete product/parts
    b) Repair product to a working order, not "as new"
    c) Remanufacture product to "as new" condition
3. Cannibalization
    a) Reuse of sub-assemblies
    b) Reuse of components
4. Material recycling
    a) Back to original application
    b) Down cycling to lower grade application
    c) Back to feedstock (revert to chemical inputs)
5. Incineration (it is not related to CLMRP activities)
    a) With energy recovery
    b) Without energy recovery
6. Disposal as waste (it is not related to CLMRP activities)
    a) Controlled
    b) Uncontrolled

A CLMRP systematic methodology can be initialized by using a software application in a production system that helps a manufacturer follow the flow of material, inventory, and completed products in the system and utilize this knowledge, in company with other production variables, to regulate operations. The concept of closed-loop manufacturing systems can incorporate all aspects of the business, including materials sourcing, production processes, sales, and the recovery processes and recycling that form the entire life cycle of a product. Some parts of recyclable, repairable, and reusable materials that can be consumed in the manufacturing process of the new products or new elements of final products are designated as inputs for a process in the system. Other parts can be sold to the related industries that need them. Waste elements that cannot be eliminated or reduced are either recaptured and biodegraded or biocomposted and sold to an industry that requests it.

A closed-loop system not only moderates the environmental impact; it also reduces current manufacturing costs, smooths raw material procurement, saves energy and water, and allows manufacturing systems in various industries to benefit from one another in cases in which one company's waste is another company's raw material or fuel. CLMRP strategies can be categorized as:

- Internal Reuse as complete product/parts
- External Reuse as complete product/parts
- Internal Repair
- External Repair
- Internal Refurbish
- External Refurbish
- Internal Remanufacture
- External Remanufacture
- Internal Recycling
- External Recycling
- Composting
- The possibility of applying biodegradation procedure

10.2.1.1. Disassembly, Remanufacturing, and Reuse

The disassembly process is a critical action in the middle and end of a product's service life cycle. Remanufacture and reuse are activities at the end of a product's service life. The foremost reason for performing these activities is to enhance sustainability, which comprises the reduction of environmental impact and increase in economic and societal benefits to meet the goal of sustainable development. An information model for disassembly, remanufacturing, and reuse can be defined and established for a manufacturing system. In addition, prototype disassembly database, disassembly/remanufacturing/or reuse cost-analysis software, and disassembly process planning systems can be developed using the information model.[7]

10.3. Multi-Criteria Decision-Making Methods for CLSC Design and Planning

A variety of multi-criteria decision-making methods containing multi-attribute decision making (MADM) such as ELECTRE, TOPSIS, ANP, AHP, and so forth and multi-objective decision making (MODM) such as compromise programming, goal programming, and so on can be applied in CLSC design and planning. For instance, in terms of choosing the best combination of suppliers based on different defined supplier selection criteria, MADM methods can be applied[8, 9] MODM methods play an essential role when multiple objectives are taken into account for a CLSC system. For instance, in a situation in which minimizing the total costs of all echelons; minimizing the carbon footprints of various facilities, sites, and transportation systems; and maximizing the service level related to final customers occur simultaneously in the given supply chain, MODM methods can be applied to solve the developed mathematical model and find the optimal solution.[10, 11]

10.4. Optimization Modelings for CLSC Design and Planning

A variety of factors such as environmental legislation, customer awareness, and the economic motivations of businesses and manufacturing systems place CLSC problems in a unique and vital subject category in terms of the classification of supply chain management issues. Designing and planning a CLSC in small and medium sizes can be done by proposing optimization models that use general exact solvers such as CPLEX, GAMS,

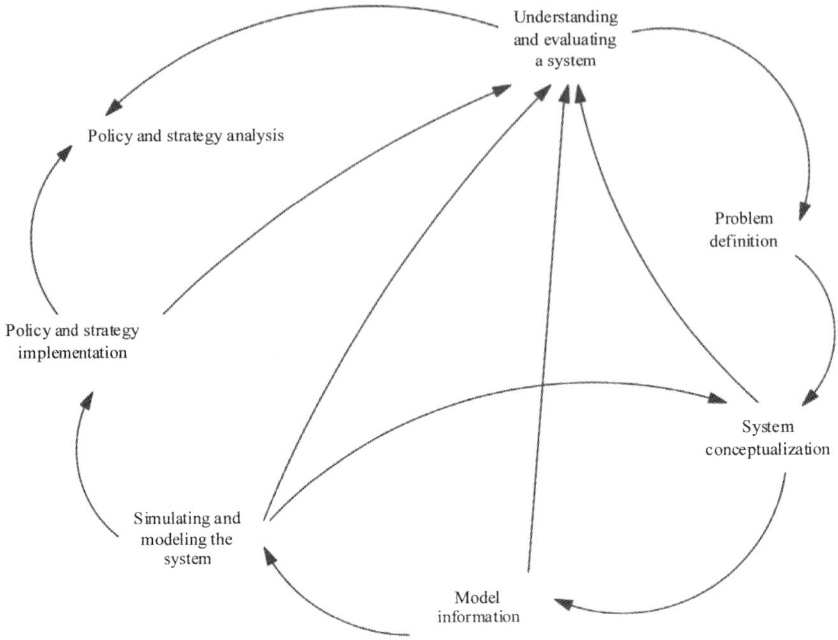

Figure 10.4. System Dynamic Modeling Process

LINDO, and LINGO to find the best scenarios and optimal solutions.[12, 13, 14, 15] Designing and planning a CLSC in very large-size instances can be taken as nondeterministic polynomial time hard (NP-hard) problem, which means that the problem can be solved in polynomial time using a nondeterministic Turing machine (like a regular Turing machine but also including a nondeterministic "choice" function). Some popular meta-heuristic algorithms to solve the mathematical optimization modeling in such cases are the genetic algorithm, particle swarm optimization algorithm, tabu search algorithm, and memetic algorithm. In some cases, proposing new and effective hybrid algorithms and methodologies can help manufacturing systems to cope with CLSC problems appropriately. It is also important to analyze the robustness of the proposed model and solution through a comprehensive evaluation study based on the considered time horizon.

10.5. System Dynamic Models and Simulation

System dynamic (SD) models are effective tools for understanding the dynamic behavior among complex systems. SD models assist the analysis to recognize the importance of variations. In a manufacturing system, an SD

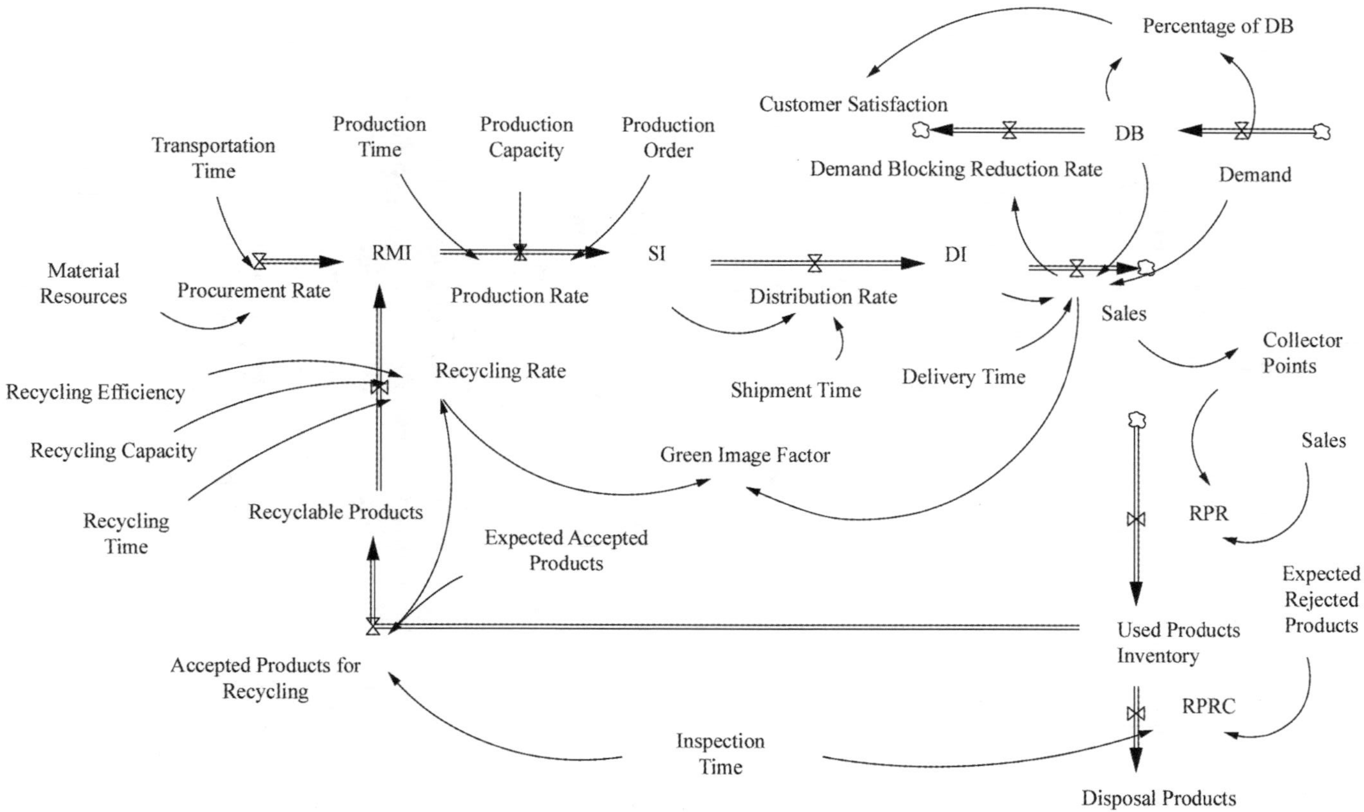

Figure 10.5. Sample of Improved Closed-Loop Supply Chain Stock and Flow Diagram Using System Dynamic Modeling

Source: Golroudbary, S. R., and Zahraee, S. M. (2015), "System dynamics model for optimizing the recycling and collection of waste material in a closed-loop supply chain."

model can be developed to investigate the implication of various elements in environmental sustainability strategies (such as environmental legislation and green image), the functional characteristics of CLSC, and the related interactions. Figure 10.4 demonstrates the general process of SD modeling.

A sample of an improved closed-loop supply chain stock and flow diagram using SD modeling presented by Golroudbary and Zahraee (2015) is shown in Figure 10.5. This flowchart contains each logical possible action in the system. In this flowchart, the terms *RMI, SI, DI, DB, RPR,* and *RPRC* refer to Raw Materials Inventory, Serviceable Inventory, Distributor's Inventory, Demand Backlog, Return Products Rate, and Rejected Products for Recycling, respectively.[16]

Qualitative analysis of dynamic behavior of the CLSC system can also be implemented by using a set of mathematical equations and simulating them by applying some popular SD software—Vensim, AnyLogic, and TRUE (Temporal Reasoning Universal Elaboration), which are effective tools for modeling, simulating, analyzing, and optimizing multidomain dynamic applications in CLSC problems. For instance, the dynamic behavior of the level variables such as Raw Materials Inventory, Serviceable Inventory, Distributor's Inventory, Used Products, and Recyclable Products can be considered by a time integral of the net inflows minus the net outflows.

10.6. Inventory Control in Closed-Loop Supply Chain

Inventory control is a central activity in CLSC design and planning, particularly for remanufacturing processes. Various inventory control models have been proposed to compare procurement and inventory control strategies and to decide on an optimized method based on cost minimization. For instance, the deterministic Economic Order Quantity (EOQ) model is one of the approaches to the inventory control system in the remanufacturing context. The EOQ model takes general policies into account as an alternation between manufacturing and remanufacturing batches. There are disposals with rate variation. In addition, in this inventory model, different holding cost rates between manufactured and remanufactured items are considered. Following the same topic, another deterministic model for an inventory system is developed where the demand is satisfied by recovering products and newly purchased products. This model uses a joint EOQ and Economic Production Quantity (EPQ) model, where one setup/batch for recovery and many orders for new purchased products, or vice versa, are concurrently taken into consideration. Some extended models such as extended EOQ analysis under stochastic production and demand rates are also presented to control inventory in CLSC systems.

Moreover, based on the nature and characteristics of the manufactured product/parts a proper inventory control review (e.g., periodic review or continuous review) can be selected. As a sample, for controlling the inventory of perishable products/parts in CLSC systems, periodic-review inventory control system is recommended.[17, 18] For these kinds of products, some analyses need to be done to coordinate inventory control and pricing strategies in a considered CLSC.

10.6.1. Inventory Optimization and Related Costs

Finding an optimal strategy for inventory optimization in a closed-loop supply chain system with remanufacturing needs some mathematical analysis. From the perspectives of the supplier, the manufacturer, the retailer, and the third-party recycling dealer/center, an optimal production and replenishment policy should be formulated to maximize the joint profit.

10.6.1.1. Example of Assumptions

Related to a mathematical analysis, some assumptions such as following ones should be determined:

- An infinite or limited planning horizon.
- No deterioration or having inventory deterioration.
- No stock shortages or shortages are allowed.
- The manufacturing and remanufacturing rates are constant parameters or they are probabilistic variables.
- Lead times are constant parameters or probabilistic variables.
- The multiechelon inventory system contains a single item or multiple items.
- The product annual demand rate and the annual return rate are constant parameters or probabilistic variables.
- The annual return rate is less than the annual demand rate.
- No space, capacity, or capital constraints or there are capacity constraints.

- No quantity discount or there are quantity discounts.
- The number of deliveries within the manufacturing cycle is an integer.
- The setup cost per run and the annual holding cost fraction are known and constant.
- To meet the retailer's demand, the remanufactured products are available before the manufactured products.
- The remanufactured products are comparable to newly manufactured products.
- Single/multi-suppliers, single/multi-manufacturers, single/multi-retailers, and single/multi–third parties in the closed-loop supply chain multiechelon inventory system are taken into consideration.

Defining the Retailer's Parameters: As a sample, some of the retailer's parameters that should be considered in the related CLSC system contain:

- Annual demand rate
- Ordering cost to the retailer
- Inventory holding cost to the retailer
- Retail price and wholesale price
- Retailer ordering cycle time (decision variable)
- Total pertinent cost to the retailer per unit of time and total retailer profit per unit of time

Defining the Manufacturer's Parameters: In addition, the manufacturer's parameters comprise the following items:

- Setup cost per manufacturing run and setup cost per remanufacturing run
- Ordering cost for the manufacturer's material warehouse and ordering cost related to the manufacturer's used product warehouse
- Product inventory holding cost to the manufacturer and material inventory holding cost for the manufacturer's warehouse
- Used product inventory holding cost of the manufacturer's warehouse
- Manufacturer unit purchase price from the supplier and manufacturer unit purchase price from the third party

- Manufacturer reproduction period / nonproduction period / production period in each remanufacturing cycle
- Number of deliveries per manufacturing/remanufacturing cycle time from the manufacturer to the retailer (decision variables)
- Total number of deliveries per remanufacturing and manufacturing cycle time from the manufacturer to the retailer (a positive integer number)
- Fixed cost to the manufacturer for processing buyer order of any size
- Annual manufacturer production rate and annual manufacturer reproduction rate
- Total pertinent cost per unit of time to the manufacturer and total relevant cost per unit of time for the material warehouse
- Total cost per unit of time for the used product warehouse and total manufacturer profit per unit of time

Defining the Third Party's Parameters: The following items refer to the third party's parameters:

- Annual return rate
- Setup cost per run for the third party
- Used product inventory holding cost to the third party
- Third party collecting the unit cost of the consumer
- Number of deliveries in manufacturer production period in each manufacturing cycle for the third party to the manufacturer (decision variable)
- Fixed third-party cost to process manufacturer orders of any size
- Total cost of third party per unit of time and total third-party profit per unit of time

Defining the Supplier's Parameters: The supplier's parameters can be considered as:

- Fixed cost related to supplier per order
- Material inventory holding cost related to the supplier, in percentage per year per dollar
- Supplier purchase unit price

- Number of deliveries per manufacturer production period in each manufacturing cycle from the supplier to the manufacturer (decision variable)
- Fixed supplier cost to process manufacturer orders of any size
- Total supplier cost per unit of time and total supplier profit per unit of time

Defining the Objective Function: The objective function can be defined as joint profit per unit of time for the whole system. Moreover, in order to develop a complete model, the inventory review system (continuous/periodic) should be determined. As a sample, Figure 10.6 represents the manufacturer-retailer inventory system based on a continuous review system. In this figure, n refers to the number of deliveries per remanufacturing cycle time from the manufacturer to the retailer (decision variable), m refers to the number of deliveries per manufacturing cycle time from the manufacturer to the retailer (decision variable), refers to manufacturer nonproduction/reproduction period in each remanufacturing cycle, g is the reorder point of retailer product inventory, and t_0 refers to its time.[19]

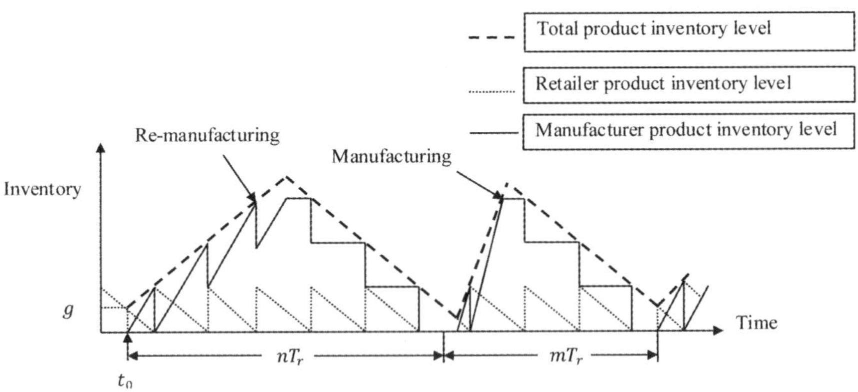

Figure 10.6. A Sample of an Integrated Manufacturer-Retailer Inventory System Based on a Continuous Review System with Remanufacturing

Source: Chung, S. L., Wee, H. M., and Yang, P. C. (2008), "Optimal policy for a closed-loop supply chain inventory system with remanufacturing."

10.7. Forecasting Methods in CLSC Systems

Forecasting techniques are powerful tools for including an estimation of demands, returns, and sales in future CLSC systems. The complexity of such techniques varies from the classical "deterministic" approach of methods, such as exponential smoothing, to the contemporary "stochastic" approach of autoregressive integrated moving average (ARIMA) models. A potential approach for obtaining returns forecasts is to apply time series models to a set of data on historical past returns. Another approach is to use dynamic regression models.[20]

References

1. Ahi, P., Searcy, C. (2013), "A comparative literature analysis of definitions for green and sustainable supply chain management," *Journal of Clean Production* 52, 329–341.

2. Sarkis, J., *Greening the Supply Chain* (London: Springer-Verlag, 2006), 251–271.

3. Handfield, R. B. (1996), "Green Supply Chain: Best Practices from the Furniture Industry," in *Proceedings of the Annual Meeting of the Decision Sciences Institute,* no.3, 1295–1297.

4. Bowen, F. E., Cousins, P. D., Lamming, R. C., and Faruk, A. C. (2001), "The Role of Supply Management Capabilities in Green Supply," *Production and Operations Management* 10, no. 2, 174–189.

5. Ramezani, M., Kimiagari, A. M., and Karimi, B. (2014), "Closed-loop supply chain network design: A financial approach," *Applied Mathematical Modelling* 38, 4099–4119.

6. Elahi, B., and Franchetti, M., "A New Optimization Model for Closed-Loop Supply Chain Networks," Technology Management Conference (ITMC), 2014 IEEE International (Chicago, June 12–15, 2014), 1–9.

7. Feng, S. C., Lee, H., Joung, C. B., Kramer, T., Ghodous, P., and Sriram, R. D. (2011), *Information Model for Disassembly for Reuse, Recycling, and Remanufacturing*, National Institute of Standards and Technology, U.S. Department of Commerce, NISTIR 7772, 3–73.

8. Elahi, B., Seyed-Hosseini, S. M., and Makui, A., "Developing a New MCDM Approach for Selecting Suppliers in a Supply Chain Under Uncertain Environment," in *Proceedings of the 6th International Congress on Logistics and SCM Systems* (International Federation of Logistics and SCM Systems and the Asian Pacific Federation of Logistics and SCM Systems, 2011), 655–667.

9. Akhavan, P., Elahi B., Jafari, M. (2014), "A new integrated knowledge model in supplier selection: The case of an Asian automotive supply chain," *Journal of Education, Business and Society: Contemporary Middle Eastern Issues* 7, no. 4, 333–368.

10. Elahi, B., Seyed-Hosseini, S. M., and Makui, A. (2011), "A fuzzy compromise programming solution for supplier selection in quantity discounts situation," *International Journal of Industrial Engineering & Production Research* 22, no. 2, 107–114.

11. Elahi, B., and Franchetti, M. (2013), "Optimization Modelling for Green Supply Chain Planning," in *Proceedings of the 106th Air and Waste Management Association*

(AWMA) Annual Conference (Pittsburgh: Air and Waste Management Association, 2012), 78.

12. Elahi, B., Seyed-Hosseini, S. M., and Makui, A. "A New Strategic Modeling Approach for Optimization of a Four-Echelon Supply Chain," in *Proceedings of the 6th International Congress on Logistics and SCM Systems* (International Federation of Logistics & SCM Systems and the Asian Pacific Federation of Logistics & SCM Systems, 2011), 281–293.

13. Elahi, B., Etaati, L., and Seyed-Hosseini, S. M., "An Integrated Modelling for Supplier Selection and Optimal Lot Sizing: A Case Study of Four-Echelon Supply Chain," Technology Management Conference (ITMC), IEEE International (San Jose, CA, June 27–30, 2011), 877–884.

14. Seyed-Hosseini, S. M., Elahi, B., and Makui, A. (2011), "Proposing a Mathematical Programming Model for Optimization of Supply Chain Scheduling," *Academy of Business Research Journal (ABRJ)* 1, 64–75.

15. Badkoobehi, H., Elahi, B., and Makui, A. (2011), "A Novel Economic Modeling for Supply Chain Planning: A Three-Echelon Supply Chain," *Business Research Yearbook* XVIII, no. 2, 433–439.

16. Golroudbary, S. R., and Zahraee, S. M. (2015), "System dynamics model for optimizing the recycling and collection of waste material in a closed-loop supply chain," *Simulation Modelling Practice and Theory* 53, 88–102.

17. Seyedhosseini, S. M., Elahi, B., and Akhlaghy, A. (2011, "Proposing a non-linear mathematical model for order splitting in a supply chain with perishable products: Solving by genetic algorithm," *International Journal of Business Research* 11, no. 4, 104–111.

18. Elahi, B., Pakzad-Jafarabadi, Y., Etaati, L., and Seyedhosseini, S. M. (2011), "Optimization of supply chain planning with considering defective rates of products in each echelon," *Technology and Investment* 2, no. 3, 211–221.

19. Chung, S. L., Wee, H. M., and Yang, P. C. (2008), "Optimal policy for a closed-loop supply chain inventory system with remanufacturing," *Mathematical and Computer Modelling* 48, 867–881.

20. Gallego, R. C., and Cueto, E. P. (2009), "Forecasting the returns in reusable containers' closed-loop supply chains: A case in the LPG industry," Proceedings of the 3rd International Conference on Industrial Engineering and Industrial Management (Terrassa, Spain, September 2–4, 2009), 311–320.

Chapter 11

Enabling and Maintaining Sustainable Manufacturing Systems

After identifying sustainability-related system improvement for a manufacturing system, enabling and maintaining these initiatives are required to ensure that the proposed benefits of the changes are realized. Proper training, communication, execution, and monitoring of the sustainability plans are critical to ensure this success.

11.1. Training

Training is a core requirement of any successful program launch, sustainability or otherwise. Training instructs employees at all levels in an organization how to implement and maintain process changes. The primary purpose of training is related to the execution of the selected sustainability strategies. The training should be focused on the employees who will be performing the new process from a managerial, technical, and shop floor level. A "just in time" training approach is usually most effective so that employees will be applying the training very shortly after learning the concepts. The goal of effective training is related to the employee's ability to perform new tasks effectively on the job and to achieve the desired goals of the changes.

The key strategy to training involves organizing and delivering material in such a way that trainees can immediately apply the concepts. On-the-job training is highly recommended, as it teaches practical application-based skills that employees will use on their jobs. For example, if a cardboard baler is installed, the operators should be given a detailed training session on the machine's operation, a standard operating procedure, a safety talk, and a list of contacts in the event the machine malfunctions or if they have improvement ideas. One-on-one coaching should also be made available as needed.

During the training session, examples of common problems and trainee exercises may be beneficial to allow the trainees to work together to solve sustainability problems in their work units and to identify new cost and quality opportunities. A group discussion can also be a useful component of the training exercise to gather feedback and employee perceptions. The trainees should be made aware of the key metrics of the sustainability program. Specifically, the environmental and economic impacts in their work units should be presented and discussed. The tracking and monitoring methods of the sustainability programs should also be discussed. Trend charts that display quality and economic performance versus expectations should be presented, as they reinforce and highlight the continued benefits of the lean six sigma programs. The location of these charts and how to read them should also be a shared vision within the manufacturing facility. This is required by all levels of management to emphasize the fit between the program and business needs. A well-thought-out vision has the power to motivate employees and ensure success of the program. Suitable locations include common areas such as lunchrooms or informational boards and should be in line with organizational policies.

11.2. Communication

Effective and continuous communication of the sustainability program is needed to successfully fulfill an organization's vision. Effective communication is an ongoing process and a key component to any sustainability program implementation plan. The four Cs of effective communication plans are: concise, consistent, complete, and creative. A strong strategy and well-thought-out plan can quickly become weak and counterproductive if they are not communicated properly to affected employees. Some methods to improve and facilitate communication within a manufacturing facility include:

- Holding regular environmental meetings with representatives from each department to discuss possible program changes on a regular basis
- Holding monthly or quarterly companywide luncheons to promote and recognize environmental efforts of the organization, departments, and individuals

- Publicizing program changes and achievements in the company newsletter
- Announcing special events in memos and informational pamphlets
- Rewarding employees for program involvement

11.3. Execution

Proper and timely execution of a well-developed deployment plan is absolutely critical to a sustainability project's success. Many well-developed strategies and programs have failed because of poor execution. There is no such thing as overcommunication when it comes to expanding or implementing a new project or program. The key components of a successful execution are creating a strong deployment plan, following the deployment plan, communication to all stakeholders, and recognizing the need to adjust in certain circumstances.

To facilitate the communication process, at a minimum weekly progress meetings should be held with all key stakeholders. These meetings should focus on the status of each project versus the timeline and established goals. An agenda and the project timeline should be prepared in advanced and serve to lead the discussion. The task leader (as determined in the deployment plan) should take the lead role in discussing the status of each project or program. Any obstacles or delays should be discussed so that the team may determine solutions.

During the deployment process it is critical not to overwhelm employees with process changes. Effort should be taken to ensure that all employees are aware of upcoming changes, timelines, and the reasons behind the change. This can be accomplished with service talks, postings, or newsletters in paycheck envelopes. All three options may be used to ensure that the message is heard, that employees are not confused, and that they buy into the programs.

In general, less effort is required for operational and process changes. These options can usually be implemented in a much quicker fashion than can equipment or material changes. A general outline of the scope of an execution plan is provided below:

- Approve the project or program
- Finalize the specifications and design for each alternative

- Submit and gather bid requests and quotes (if necessary)
- Complete and submit a purchase order
- Receive and install the equipment
- Finalize operating and maintenance procedures
- Train affected employees
- Start the project or program
- Complete regulatory inspections
- Track implemented project cost savings and waste reductions

11.4. Monitoring and Control

After the sustainability program has been implemented, it should be monitored on a periodic basis to ensure that it is still performing as planned and to make any necessary adjustments. This includes monitoring solid waste and air emissions, employee behavior, operational performance, and financial performance versus the goals. In addition, emphasis should be placed on continuous improvement to enhance current sustainability programs and to identify new opportunities. It may be beneficial to conduct period assessments, facility walk-throughs, or employee interviews by the original waste reduction team to accomplish these goals. When evaluating the program, it is important to:

- Keep track of program success and build on past successes
- Identify new ideas for waste reduction
- Identify areas needing improvement
- Document compliance with state or local regulations
- Determine the effect of new additions to the facility or program
- Keep employees informed and motivated

In addition, consider reviewing the organization's environmental and purchasing records on at least a quarterly basis to ensure that the sustainability program is still in effect and performing as expected. New product or process changes should also be evaluated at the onset to ensure that the design minimizes environmental impact. This is easily accomplished by adding "sustainability review" to the new product or process checklist or standard operating procedure.

The sustainability program is a continuing project, not a onetime event. The goal is to integrate sustainability into the "culture" of the organization. Generally, the sustainability assessment and implemented alternatives will target only the high need areas. Once these high need areas have been addressed, the team can focus on other areas in the facility. From a systems standpoint, the ultimate goal of the team is to minimize the environmental footprint of the facility. The frequency in which the additional assessments are conducted will depend on the budget of the company. In general, organizations that conduct assessments one to four times per year have achieved paybacks. In addition, if there are special circumstances that indicate the need for further review, an assessment should be conducted. These special circumstances include:

- A change in raw material or product requirements
- Higher waste management costs
- New regulations
- New technology
- A major event with undesirable environmental consequences (e.g., a major spill)

To be truly effective, an organizational culture of sustainability must be fostered within the organization. Executive management must ensure this through repeated communications and acknowledgments of success stories from individuals or business units. This will make sustainability an integral part of the organization's manufacturing operations.

Chapter 12

Case Study: An American Multinational Power Management Company

The company case study is an American multinational power management company with its head office in Dublin, Ireland (its operational headquarters are located in the United States, in Ohio), providing solutions to its customers to manage electrical, hydraulic, and mechanical power efficiently, safely, and sustainably. This company, which is a public company and has been categorized as a conglomerate industry, was founded in 1911 in Bloomfield, New Jersey, and in 1914 moved to Cleveland, Ohio, to be closer to its core business, the automotive industry.

This company acquired Cooper Industries in November 2012. It has various worldwide locations such as the United States, Asia, the United Kingdom, Germany, France, and Mexico. The 2013 revenue of the combined companies was about $22 billion.

This company comprises two main sectors. The Electrical Sector and Industrial Sector. The Electrical Sector is one of the global leaders in electrical products, systems, and services for power quality, distribution and control, power transmission, and lighting and wiring products. Products include electrical and industrial components, residential products, single- and three-phase power quality, emergency lighting, fire detection, wiring devices, structural support systems, circuit protection, lighting products, power distribution and assemblies, hazardous duty electrical equipment, intrinsically safe explosion-proof instrumentation, utility power distribution, power reliability equipment, and services. The Industrial Sector contains three main segments: Hydraulics Segment, Aerospace Segment, and Vehicle Segment.

12.1. Environmental Record

Since 2011, this company has employed effective designs for the environment principles to reduce the environmental impact of company products across their life cycles. In 2013, this company was ranked a global leader by the Carbon Disclosure Project (CDP).

Regarding the key environment, health, and safety metrics defined by the company in 2013, greenhouse gas (GHG) emissions were reduced by 2.8% on an absolute basis. The use of virgin polystyrene product packaging was reduced by 32%, or nearly 239,059 kilograms (527,026 pounds)—the equivalent of eliminating 527 dumpsters of waste from landfills. Water consumption was decreased by 4.5% in 2013, when compared with 2012. Moreover, 75% of its key suppliers agreed to provide emissions, energy use, and water consumption data to the CDP supply chain program.

This company has defined its environmental goals for 2014–2015 (on both an absolute and indexed basis) as follows:

- To reduce GHG emissions by 3%
- To reduce waste to landfill by 3%
- To reduce water consumption by 5%

Moreover, this company reports Environmental Protection Agency (EPA) Toxics Release Inventory (TRI) data from all U.S. facilities annually. The TRI is a public database containing information on disposal and other releases of over 600 toxic chemicals from thousands of U.S. facilities. Information about how facilities manage those chemicals through recycling, energy recovery, and treatment is also provided.

Each year, companies across a wide range of industries (including chemical, mining, paper, oil, industrial, and gas industries) that produce more than 11,340 kilograms (25,000 pounds) or handle more than 4,536 kilograms (10,000 pounds) of a listed toxic chemical must report it to the TRI. The data are then posted on the EPA's TRI web page.[1]

12.2. Methodology

Before performing any assessment, it is essential to define a proper methodology, set achievable objectives and goals, and choose appropriate strategies. In the following two sections, we focus on the methodology applied for data collection and assessment regarding solid and hazardous waste and energy.

12.2.1. Solid and Hazardous Waste

The solid waste assessment team targeted one of the operational headquarters of this company in Ohio and conducted a solid and hazardous waste assessment at that facility. The studied company had an existing recycling program for metals, computer waste, cardboard, and batteries, and at the time of the audit, it was spending approximately $4,200 for waste disposal per year. The objectives of this recycling research were to determine additional recycling opportunities at this company and to explore reducing the cost of waste disposal for the facility. The objectives included identifying the major solid waste streams and suggesting potential areas for reducing, reusing, and recycling materials in the waste stream. Related to solid waste assessment and recycling processes, the ISO standard (ISO, 2006) also recommends that the environmental benefits of recovered resources should be accounted for by broadening the system boundaries to include the avoided burdens of conventional production.[2]

Moreover, the recycling (%) and the biological treatment (%), which are the percentages of material that have been recycled or biologically treated, can be calculated according to the following equations (Eq.1 and Eq.2):[3]

$$\text{Recycling rate} = \frac{\text{Recycled material}}{\text{Material available for recycling}} \tag{1}$$

$$\text{Biological treatment rate} = \frac{\text{Biologically treated material}}{\text{Material available for biological treatment}} \tag{2}$$

The procedure for this assessment included examining each of the waste containers in the facility and inspecting the current recycling programs. The containers were each inspected to determine the total volumes of waste, types of materials, and quantity of each material. From this information, estimated volumes of each material were calculated for each waste container.[4] Using standard packing densities, weights of recyclable materials were estimated (Table 12.1). Volumes are required to determine the waste hauling cost savings, and weights are required to determine the revenue generated from the sale of recyclables. Using the data collected, annual waste streams were calculated.

12.2.2. Energy

A project was defined to determine alternatives that would reduce energy and reduce costs for the considered company. The objectives of this project were to define the energy consumption at this company and to identify economically feasible options for energy reduction. During the visit, an energy assessment was conducted and energy consumption data were obtained from related managers and staffs. The energy assessment consisted of a comprehensive examination of current lighting fixtures and appliances in the entryway and conference rooms and in the office, manufacturing, and kitchenette areas. Energy minimization options for the major areas of energy consumption were researched with an emphasis placed on energy reduction and economic feasibility.

Table 12.1. Standard Packing Densities for Different Kinds of Materials in a Solid Waste Stream

Waste	Density (lbs./cubic yard)
Mixed office paper (MOP)	110
Shredded mixed office paper (S.MOP)	70
Old newspaper (ONP)	157.50
Magazine	205
Aluminum cans	32
Old corrugated cardboards (OCC)	50
Mixed solid waste (MSW)	53
PET (1)	40
HDPE (2)	1,601.28
Glass	600

12.3. Solid Waste Reductions

During the assessment at this company the waste receptacles were examined and data were gathered and analyzed to calculate the weight of major solid wastes disposed of annually by the company.

Table 12.2. Annual Weight of Solid Waste Stream and Disposal Data

Component	Total weight (tons/year)	% of total weight	Waste recycled (tons/year)	Waste not recycled (tons/year)
Waste Streams That Can Be Recycled				
Mixed office paper	2.0	15.2	0.0	2.0
Newspaper	0.9	6.5	0.0	0.9
Magazines	1.9	14.5	0.0	1.9
Cardboard	2.1	15.4	1.1	1.0
Aluminum cans	0.5	3.8	0.0	0.5
PET (1)	0.6	4.5	0.0	0.6
HDPE (2)	0.6	4.2	0.0	0.6
Other plastics	1.0	7.7	0.0	1.0
Plastic wrap	0.0	0.1	0.0	0.0
Subtotal	9.7	71.8	1.1	8.6
Waste Streams That Cannot Be Recycled				
Nonrecyclable/ Food	3.3	24.3	0.0	3.3
Other	0.5	3.9	0.0	0.5
Subtotal	3.8	28.2	0.0	3.8
Grand total	13.5	100.0	1.1	12.4

Table 12.2 and Figure 12.1 display the weight of major solid wastes disposed of annually by the company. The waste streams are shown by annual tonnage and by the percentage of total waste. As shown in Table 12.2, the facility currently disposes of approximately 13.5 tons of solid waste each year. Of that total it was determined that 1.1 tons are currently being recycled, 8.6 tons are recyclable but not currently being recycled, and 3.8 tons are wastes that are not recyclable in the common markets.

Furthermore, the assessment shows that 402.4 Yd^3 of potentially recyclable material is thrown away each year (Table 12.3 and Figure 12.2). Utilizing the provided waste hauling cost of $837.13 for every 30 Yd^3 container, an annual cost savings was determined. The recycling of all potentially recyclable materials would result in an annual cost savings of $11,228.70. Please note that, this is a calculated cost savings using conservative values and also does not take into account savings related to the potential sale of recycled materials.

The aforementioned results indicate that there are several waste streams that have the potential to be recycled. These streams include mixed

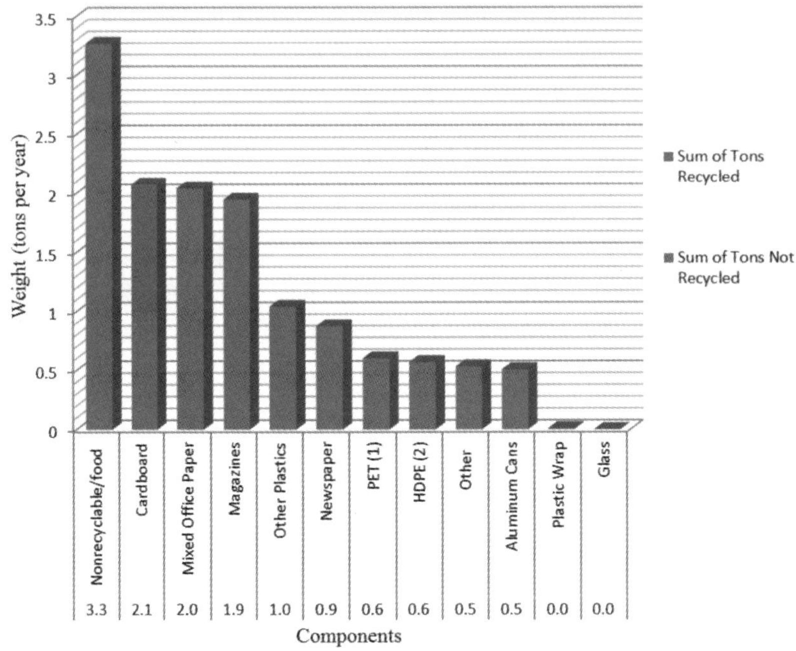

Figure 12.1. Annual Weight of Major Solid Waste Streams

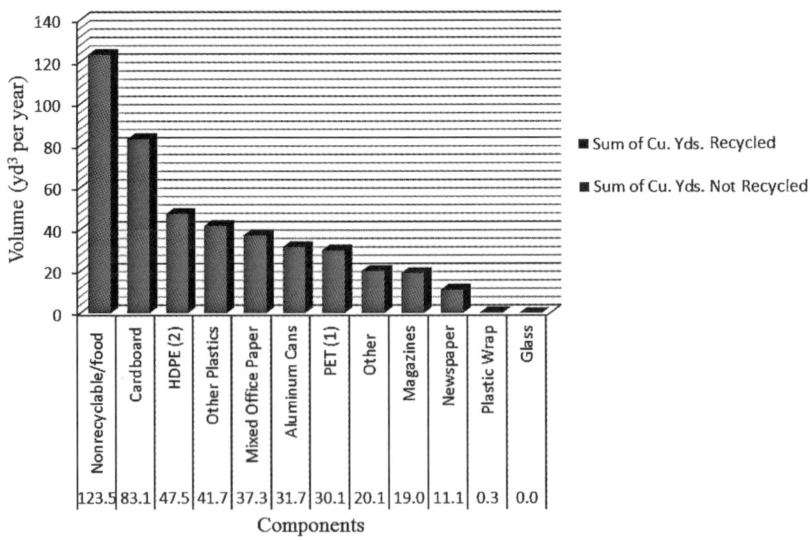

Figure 12.2. Annual Volume of Major Solid Waste Streams

Table 12.3. Annual Volume of Major Solid Waste Streams

Component	Total volume (yd^3/year)	% of total volume	Waste recycled (yd^3/year)	Waste not recycled (yd^3/year)
Waste Streams That Can Be Recycled				
Mixed office paper	37.3	8.4	0.0	37.3
Newspaper	11.1	2.5	0.0	11.1
Magazines	19.0	4.3	0.0	19.0
Cardboard	83.1	18.7	43.0	40.1
Aluminum cans	31.7	7.1	0.0	31.7
PET (1)	30.1	6.8	0.0	30.1
HDPE (2)	47.5	10.7	0.0	47.5
Other plastics	41.7	9.4	0.0	41.7
Plastic wrap	0.3	0.1	0.0	0.3
Subtotal	301.8	67.8	43.0	258.8
Waste Streams That Cannot Be Recycled				
Food waste	123.5	27.7	0.0	123.5
Other	20.1	4.5	0.0	20.1
Subtotal	143.6	32.2	0.0	143.6
Grand total	445.3	100.0	43.0	402.4

office paper, cardboard, plastics, magazines, and newspapers. It should also be noted that the studied company recycles all of the aluminium, brass, and steel scraps, which is an effective current process. And there is a notable recycling effort for batteries and the waste of electronic equipment.

12.3.1. Reduction in Waste Hauling Costs

The above charts and tables outline the major annual waste streams at this company. These charts identify both the recyclable and nonrecyclable elements in each stream. Placing clearly marked recycling containers near existing waste containers will help to increase the use of such recycling containers. This will also make it more convenient for janitorial staff to collect the recyclables.

Cost savings could be realized from a reduction in waste hauling costs. Based on the assessments conducted by the waste assessment team, waste hauling costs can be reduced by approximately 45% by reducing the volume of waste taken by the hauler and the frequency of solid waste collections. This corresponds to an annual waste hauling savings of $1,890. Many waste hauling companies can be contracted to pick up recyclable materials.

12.3.2. Sale of Recyclable Materials

Sale of recyclable materials can be considered another opportunity in the recycling program. Using average prices for selling recyclable materials based on current market, savings can be realized from the sale of materials at this company (Table 12.4). Based on computed cost savings in the previous section (12.4.1) and evaluated savings shown in Table 12.4, the total combined annual economic benefit from decreased disposal costs and increased revenue is $3,205 ($1,890 + $1,315 = $3,205).

It is worth noting, however, that three assumptions were made in Table 12.4. The first assumption is that the material is all baled before sale. The prices will be significantly lower if the material is not baled, and many companies will not purchase nonbaled materials.

The second assumption is that all of the potentially recyclable materials are captured. This is not an entirely practical assumption, but the actual amount that will be is extremely hard to determine and depends on the placement of recycling containers, the display of the containers, and employee awareness of the program. The number listed above is the goal to aim for, but it should not be used for planning decisions.

This analysis also shows that this company pays the waste hauler for each time that the waste receptacle is hauled. It currently pays for the removal of 46 tons per year and from considering the conducted audit, we can estimate that the total waste generation of the facility is about 14 tons per year. Considering this matter, if this company were to negotiate its contract with the current waste hauler, the pickups could be reduced from three times a week to two times a week. This reduction in picking up the waste could potentially save the company about $1,400 per year in waste removal fees based on a usual rate.

Table 12.4. Revenue Generated from the Sale of Recyclable Material

Component	Price/ton	Tons diverted	Revenue generated
Paper	$45	2	$90
Newspaper	$50	0.9	$45
Old corrugated cardboard	$70	1	$70
Aluminum cans	$2,000	0.5	$1,000
Plastic	$50	2.2	$110
Total		6.6	$1,315

Table 12.5. Solid Waste Minimization Recommendations

Item	Recommendation
Improving the recycling program with the addition of separated recycling containers placed strategically throughout the facility	Expanding the program of recycling to include containers for collecting paper and plastic products along with aluminium cans. This can be done by adding separated containers for recycling near the current waste receptacles. It is better to place containers in locations that are more visible.
Reducing the frequency of waste hauling	Savings are possible by reducing the frequency of times the solid waste at the facilities is collected. Oftentimes, the waste hauling companies are on a fixed hauling schedule and end up emptying the dumpsters when they are half full or even less than that.
Increasing the capture rate of cardboard in the current program.	Use employee awareness to help capture more of the cardboard waste stream that is slipping through the current system. This can be done with a memo about the program, or it could even be brought up briefly in an already scheduled meeting.
Sale of recyclable materials	Some kinds of recyclable materials can be sold in related markets. In this way, recyclable materials can generate revenue rather than being considered as disposals.
Establishing more effective education programs	The goal is to educate both customers and employees in the importance of recycling. Placing in every room handouts and hanging posters and flyers describing what can be recycled and where to recycle it can be effective methods.
Utilizing composting to reduce food waste	Composting food waste would have two benefits: The amount of food waste that is sent to landfills could be reduced, and an organic fertilizer could be created for use in the garden. Food was the largest nonrecyclable waste that was identified by the audit, and composting would help to reduce this waste stream.
Food waste to be sent to an energy reclamation facility	This method helps to reduce the amount of waste going to landfills. A complete benefit-cost analysis should be conducted to evaluate shipping costs associated with this method.

12.4. Solid Waste Minimization Recommendations

Table 12.5 provides a list of recommendations for this company to minimize solid waste disposal. Each item and a detailed description are provided. As it is shown in this table, "composting" and "sending food waste to an energy reclamation facility" have been recommended as some proposed methods to reduce food waste. Recently lots of research has focused on economic analysis regarding reduction of food waste, especially in European countries. *Government Review of Waste Policy in England* in *2011* states that "compared to land filling, each ton of food waste prevented means that 4.2 tonnes of CO_2 equivalent [CO_2 equivalent is the concentration of CO_2. It is expressed as parts per million by volume] emissions are avoided, while only around 500 kilograms is avoided for each tonne processed through AD [anaerobic digestion], the best currently available treatment option, and even less for other options such as composting or energy recovery."[5, 6]

12.5. Data Analysis and Energy Consumption Assessment

In terms of energy assessment, complete examinations of current lighting fixtures and appliances in various areas of the facility were conducted. These areas contain the entryway, conference rooms, office area, manufacturing sections, and kitchenette area.

Data Analysis of Office Area: Throughout the office area was standard fluorescent lighting, which is one of the top contributors of energy usage. The other factors were personal computers (mostly laptops) and the printers throughout the office area. The team's focus was put on the lighting, as it is the major energy consumer that can be reduced. Note that further energy savings can be found through the employees' computers. Throughout all the offices, the team found the information shown in Table 12.6. Considering Table 12.6, the letters *A, C,* and *F* were used only as a notation for the team's observation purposes. The same letters are also used in Table 12.7 to evaluate the energy consumption and total cost of the lighting for

Table 12.6. Office Lighting Breakdown by Fixture Type, Quantity, and Wattage Used in Energy Assessment

Fixture type	Fixture details	Wattage	Quantity	Total wattage in kW = (wattage × quantity) /1000
A	4-ft fluorescent lights	40	74	2.96 kW
C	*U*-shaped fluorescent lights	40	551	22.04 kW
F	4-ft fluorescent lights	32	6	0.19 kW

each time period (year, month, week, day, and hour). Note that costs were estimated at $0.095 per kilowatt hour.

Data Analysis of Entryway Area: Throughout the entryway was a mix of standard fluorescent lighting and recessed lighting. There were also two TVs, which were normally powered on, showing a slide show about the company. Energy usage of these three objects can be reduced with using other kinds of fixtures or finding efficient ways to reduce their powered-on hours.

Table 12.7. Evaluation of Energy Consumption per Time Period (Hour, Day, Week, Month, and Year) and Total Cost of Energy Consumption Related to the Lighting of Office Area

Fixture type	Energy consumption per time period per unit of each fixture type in kWh		Energy consumption per time period per all units of each fixture type in kWh	Cost of energy consumption related to office lighting per time period per all units of each fixture type in $
A	Hourly	0.04	0.04 × 74 = 2.96	2.96 × 0.095 = 0.28
	Daily	0.4	0.4 × 74 = 29.6	29.6 × 0.095 = 2.81
	Weekly	1.98	1.98 × 74 = 146.52	146.52 × 0.095 = 13.92
	Monthly	8.27	8.27 × 74 = 611.98	611.98 × 0.095 = 58.14
	Annually	99.19	99.19 × 74 = 7,340.00	7340.00 × 0.095 = 697.30
C	Hourly	0.04	0.04 × 551 = 22.04	22.04 × 0.095 = 2.09
	Daily	0.48	0.48 × 551 = 264.48	264.48 × 0.095 = 25.13
	Weekly	2.40	2.40 × 551 = 1,322.40	1322.40 × 0.095 = 125.63
	Monthly	10.00	10.00 × 551 = 5,510.00	5510.00 × 0.095 = 523.45
	Annually	120.00	120.00 × 551 = 66,120.00	66120.00 × 0.095 = 6,281.40
F	Hourly	0.03	0.03 × 6 = 0.18	0.18 × 0.095 = 0.02
	Daily	0.38	0.38 × 6 = 2.28	2.28 × 0.095 = 0.22
	Weekly	1.92	1.92 × 6 = 11.52	11.52 × 0.095 = 1.09
	Monthly	8.00	8.00 × 6 = 48.00	48.00 × 0.095 = 4.56
	Annually	96.00	96.00 × 6 = 576.00	567.00 × 0.095 = 53.86
Total cost of energy consumption related to office lighting per time period in $			Hourly	0.28 + 2.09 + 0.02 = 2.39
			Daily	2.81 + 25.13 + 0.22 = 28.16
			Weekly	13.92 + 125.63 + 1.09 = 140.64
			Monthly	58.14 + 523.45 + 4.56 = 586.15
			Annually	697.30 + 6,281.40 + 53.86 = 7,032.56

Table 12.8. Entryway Light Breakdown by Fixture Type, Quantity, and Wattage Used in Energy Assessment

Fixture type	Fixture details	Wattage	Quantity	Total wattage in kW = (wattage × quantity)/1000
A	4-in fluorescent lights	40	14	0.56 kW
B	Incandescent lights	135	4	0.54 kW

The team found the information listed in Table 12.8 regarding the lighting fixtures. Considering Table 12.8, the letters were used only as a notation for the team's observation purposes. The same letters are also used in Table 12.9 to evaluate the energy consumption and total cost of the light-

Table 12.9. Evaluation of Energy Consumption per Time Period (Hour, Day, Week, Month, and Year) and Total Cost of Energy Consumption Related to the Lighting of Entryway Area

Fixture type	Energy consumption per time period per unit of each fixture type in kWh		Energy consumption per time period per all units of each fixture type in kWh	Cost of energy consumption related to entryway lighting per time period per all units of each fixture type in $
A	Hourly	0.04	0.04 × 14 = 0.56	0.56 × 0.095 = 0.05
	Daily	0.48	0.48 × 14 = 6.72	6.72 × 0.095 = 0.64
	Weekly	2.40	2.40×14 = 33.60	33.60 × 0.095 = 3.19
	Monthly	10.00	10.00 × 14 = 140.00	140.00 × 0.095 = 13.30
	Annually	120.00	120.00 × 14 = 1,680.00	1,680.00 × 0.095 = 159.6
B	Hourly	0.13	0.13 × 4 = 0.52	0.52 × 0.095 = 0.05
	Daily	1.61	1.61 × 4 = 6.44	6.44 × 0.095 = 0.61
	Weekly	8.04	8.04 × 4 = 32.16	32.16 × 0.095 = 3.05
	Monthly	33.50	33.50 × 4 = 134.00	134.00 × 0.095 = 12.73
	Annually	402.00	402.00 × 4 = 1,608.00	1,608.00 × 0.095 = 152.76
Total cost of energy consumption related to entryway lighting per time period in $			Hourly	0.05 + 0.05 = 0.10
			Daily	0.64 + 0.61 = 1.25
			Weekly	3.19 + 3.05 = 6.24
			Monthly	13.3 + 12.73 = 26.03
			Annually	159.6 + 152.76 = 312.36

Table 12.10. Evaluation of Energy Consumption per Time Period (Hour, Day, Week, Month, and Year) and Total Cost of Energy Consumption Related to TV Sets Located in Entryway Area

TV set type	Energy consumption per time period per unit of each TV set type in kWh		Energy consumption per time period per all units of each TV set type in kWh	Cost of energy consumption related to each type of TV set located in entryway area per time period in $
37-in LG	Hourly	0.18	0.18 × 1 = 0.18	0.18 × 0.095 = 0.017
	Daily	2.16	2.16 × 1 = 2.16	2.16 × 0.095 = 0.205
	Weekly	10.80	10.80 × 1 = 10.80	10.80 × 0.095 = 1.03
	Monthly	35.00	35.00 × 1 = 35.00	35.00× 0.095 = 3.32
	Annually	420.00	420.00 × 1 = 420.00	420.00 × 0.095 = 39.90
42-in Tyco	Hourly	540.00	0.18 × 1 = 0.18	0.18 × 0.095 = 0.017
	Daily	0.18	2.16 × 1 = 2.16	2.16 × 0.095 = 0.205
	Weekly	2.16	10.80 × 1 = 10.80	10.80 × 0.095 = 1.03
	Monthly	10.80	45.00 × 1 = 45.00	45.00 × 0.095 = 4.27
	Annually	45.00	540.00 × 1 = 540.00	540.00 × 0.095 = 51.3
Total cost of energy consumption related to TV sets located in entryway area per time period in $			Hourly	0.017 + 0.017 = 0.034
			Daily	0.205 + 0.205 = 0.41
			Weekly	1.03 + 1.03 = 2.06
			Monthly	3.32 + 4.27 = 7.59
			Annually	39.9 + 51.3 = 91.2

ing for each time period (year, month, week, day, and hour). Note that costs were estimated at $0.095 per kilowatt hour.

Moreover, the team evaluated the energy consumption of television sets located in the entryway area of the facility (Table 12.10). The TVs that were in use were a 37-inch (94-cm) LG LCD set and a 42-inch (107-cm) Tyco Electronics set. It is assumed that both sets are on for 60 hours per week. Costs were estimated at $0.095 per kilowatt hour.

It should be noted that the data in Table 12.10 do not reflect the cost of the TVs' standby power, or the power consumed while the set is plugged in but turned off. The specifications found this to consume < 1 watt of power.

Data Analysis of Kitchenette Area: The kitchenette area contained many high-energy appliances, including a dishwasher, a coffee machine, two different types of microwave ovens, a refrigerator, and two different types of vending machines. Each of these appliances consumes a large amount of energy. Standby power is also consumed when the appliance is not running, for instance, coffee machine and microwave ovens. Table 12.11 shows the standard average values of power for two modes of "Powered-on" and "Standby," estimated energy consumption, and total costs per time unit regarding each appliance.[7] Costs were estimated at $0.095 per kilowatt hour.

Related to the data analysis in Table12.11, please note that some of the cost estimates may be overstated. This is because it is assumed the machines are running at full power for the duration of the estimated time. Appliances tend to cycle between standby and powered modes and are difficult to estimate.

Data Analysis of Manufacturing Area: Standard fluorescent lighting was found throughout the manufacturing area. Considering that, the team gathered the information listed in Table 12.12. In Table 12.12, the letters were used only as a notation for the team's observation purposes. The same letters are also used in Table 12.13 to evaluate the energy consumption and total cost of the lighting for each time period (year, month, week, day, and hour). Note that costs were estimated at $0.095 per kilowatt hour.

Table 12.11. Evaluation of Energy Consumption per Time Period (Day, Month, and Year) and Total Cost of Energy Consumption Related to Kitchenette Appliances Located in Kitchenette Area

Appliance	Wattage of powered-on mode	Wattage of standby mode (estimation)	Energy consumption per time period per unit of each appliance in kWh		Cost of energy consumption related to each appliance located in kitchenette per time period in $
Dishwasher	1,200	N/A	Daily	2.41	2.41 × 0.095 = 0.23
			Monthly	20.3	20.3 × 0.095 = 1.93
			Annually	244.0	244.0 × 0.095 = 23.18
Ice maker	840	N/A	Daily	20.2	20.2 × 0.095 = 1.92
			Monthly	611.5	611.5 × 0.095 = 58.09
			Annually	7,338.2	7338.2 × 0.095 = 697.129
Samsung microwave oven	1,200	3	Daily	2.47	2.47 × 0.095 = 0.23
			Monthly	52.0	52.0 × 0.095 = 4.94
			Annually	624.0	624.0 × 0.095 = 59.28
Sharp microwave oven	1,100	3	Daily	2.27	2.27 × 0.095 = 0.21
			Monthly	47.8	47.8 × 0.095 = 4.54
			Annually	574.0	574.0 × 0.095 = 54.53
Refrigerator	1,800*	N/A	Daily	1.5	1.5 × 0.095 = 0.14
			Monthly	45.8	45.8 × 0.095 = 4.35
			Annually	550.0	550.0 × 0.095 = 52.25
Vending machine	200	N/A	Daily	4.8	4.8 × 0.095 = 0.45
			Monthly	145.6	145.6 × 0.095 = 13.83
			Annually	1,747.2	1747.2 × 0.095 = 165.98
Vending machine (cold)	400	N/A	Daily	9.6	9.6 × 0.095 = 0.91
			Monthly	291.2	291.2 × 0.095 = 27.66
			Annually	3,494.4	3,494.4 × 0.095 = 331.96
Coffee machine	1,800	250	Daily	12.2	12.2 × 0.095 = 1.16
			Monthly	301.7	301.7 × 0.095 = 28.66
			Annually	3,620.0	3,620.0 × 0.095 = 343.9
Total cost of energy consumption related to all appliances located in kitchenette area per time period in $				Daily	4.74
				Monthly	144.00
				Annually	1,728.21
Total energy consumption related to all appliances located in kitchenette area per time period in kWh				Daily	55.45
				Monthly	1,515.90
				Annually	18,191.80
**The refrigerator usage was estimated with online estimates of an average annual usage at 550 kWh. The wattage was not used in any calculations.*					

Table 12.12. Breakdown of Lights in Manufacturing Area Based on Fixture Type, Quantity, and Wattage Used in Energy Assessment

Fixture type	Fixture details	Wattage	Quantity	Total wattage in kW = (wattage × quantity)/1000
D	*U*-shaped fluorescent lights	40	224	8.96 kW
E	8-in fluorescent lights	60	214	12.84 kW
F	4-in fluorescent lights	32	26	0.83 kW

Table 12.13. Evaluation of Energy Consumption per Time Period (Hour, Day, Week, Month, and Year) and Total Cost of Energy Consumption Related to the Lighting of Manufacturing Area

Fixture type	Energy consumption per time period per unit of each fixture type in kWh		Energy consumption per time period per all units of each fixture type in kWh	Cost of energy consumption related to manufacturing lighting per time period per all units of each fixture type in $
D	Hourly	0.04	0.04 × 224 = 8.96	8.96 × 0.095 = 0.85
	Daily	0.96	0.96 × 224 = 215.04	215.04 × 0.095 =20.43
	Weekly	4.80	4.80 × 224 = 1,075.20	1,075.20 × 0.095 = 102.14
	Monthly	20.00	20.00 × 224 = 4,480.00	4,480.00 × 0.095 = 425.60
	Annually	240.00	240.00 × 224 = 53,760.00	53,760.00 × 0.095 = 5,107.2
E	Hourly	0.06	0.06 × 214 = 12.84	12.84 × 0.095 = 1.22
	Daily	1.44	1.44 × 214 = 308.16	308.16 × 0.095 = 29.27
	Weekly	7.20	7.20 × 214 = 1,540.80	1,540.80 × 0.095 = 146.38
	Monthly	30.00	30.00 × 214 = 6420.00	6,420.00 × 0.095 = 6,09.9
	Annually	360.00	360 × 214 = 77,040.00	77,040.00 × 0.095 = 7,318.8
F	Hourly	0.03	0.03 × 26 = 0.83	0.83 × 0.095 = 0.08
	Daily	0.77	0.77 × 26 = 19.97	19.97 × 0.095 = 1.90
	Weekly	3.84	3.84 × 26 = 99.84	99.84 × 0.095 = 9.48
	Monthly	16.00	16.00 × 26 = 416.00	416.00 × 0.095 = 39.52
	Annually	192.00	192.00 × 26 = 4,992.00	4,992.00 × 0.095 = 474.24
Total cost of energy consumption related to manufacturing lighting per time period in $			Hourly	2.15
			Daily	51.60
			Weekly	258.00
			Monthly	1,075.02
			Annually	12,900.24

Data Analysis of Other Areas Containing Restrooms, Hallways, and Conference Rooms: Throughout the other areas such as restrooms, hallways, and conference rooms, the data are shown in Table 12.4. Considering Table 12.14, the letters were used only as a notation for the team's observation purposes. The same letters are also used in Table 12.15 to evaluate the energy consumption and total cost of the lighting for each time period (year, month, week, day, and hour). Note that costs were estimated at $0.095 per kilowatt hour.

12.6. Cost-Saving Opportunities/Recommendations Related to Energy Consumption

In this section, we shed light on various cost-saving opportunities based on the gathered data regarding each area of the facility.

Table 12.14. Breakdown of Lights in Other Areas Such as Restrooms, Hallways, and Conference Rooms Based on Fixture Type, Quantity, and Wattage Used in Energy Assessment

Fixture type	Fixture details	Wattage	Quantity	Total wattage in kW = (wattage × quantity)/1000
B	Incandescent lights	135	39	5.265 kW
D	*U*-shaped fluorescent lights	40	24	0.960 kW
F	4-ft fluorescent lights	32	24	0.768 kW

Table 12.15. Evaluation of Energy Consumption per Time Period (Hour, Day, Week, Month, and Year) and Total Cost of Energy Consumption Related to Other Areas Such as Restrooms, Hallways, and Conference Rooms

Fixture type	Energy consumption per time period per unit of each fixture type in kWh		Energy consumption per time period per all units of each fixture type in kWh	Cost of energy consumption per time period per all units of each fixture type in $
B	Hourly	0.06	0.06 × 39 = 2.39	2.39 × 0.095 = 0.23
	Daily	0.73	0.73 × 39 = 28.62	28.62 × 0.095 = 2.72
	Weekly	3.67	3.67 × 39 = 143.10	143.10 × 0.095 = 13.59
	Monthly	15.29	15.29 × 39 = 596.25	596.25 × 0.095 = 56.64
	Annually	183.46	183.46 × 39 = 7,155.00	7,155.00 × 0.095 = 679.72
D	Hourly	0.04	0.04 × 24 = 1.04	1.04 × 0.095 = 0.098
	Daily	0.52	0.52 × 24 = 12.48	12.48 × 0.095 = 1.18
	Weekly	2.60	2.60 × 24 = 62.40	62.40 × 0.095 = 5.93
	Monthly	10.83	10.83 × 24 = 260.00	260.00 × 0.095 = 24.7
	Annually	130.00	130.00 × 24 = 3,120.00	3,120.00 × 0.095 = 296.4
F	Hourly	0.04	0.04 × 24 = 0.90	0.90 × 0.095 = 0.08
	Daily	0.45	0.45 × 24 = 10.75	10.75 × 0.095 = 1.02
	Weekly	2.24	2.24 × 24 = 53.76	53.76 × 0.095) = 5.11
	Monthly	9.33	9.33 × 24 = 224.00	224.00 × 0.095) = 21.28
	Annually	112.00	112.00 × 24 = 2,688.00	2,688.00 × 0.095) = 255.36
Total cost of energy consumption per time period in $			Hourly	0.410
			Daily	102.62
			Weekly	24.63
			Monthly	4.93
			Annually	1,231.49

Cost-Saving Opportunities for Energy Consumption in Office Area: While the lights used in the office area are pretty efficient, with the emergence of light-emitting diode (LED) lighting, there is some room for improvement, especially for the long term. Companies are manufacturing

Table 12.16. Evaluation of Energy Consumption per Time Period (Hour, Day, Week, Month, and Year) and Total Cost of Energy Consumption Related to Office Light Replacement

Time period	Energy consumption (kWh)		Cost of energy consumption in $
	Per unit of each fixture	For the whole office area	
Hourly	0.017	0.017 × ((74+551+6) = 630) = 10.7	10.7 × 0.095 = 1.02
Daily	0.200	0.200 × 630 = 126.11	126.11 × 0.095 = 11.98
Weekly	0.998	0.998 × 630 = 629.6	629.6 × 0.095 = 59.81
Monthly	4.16	4.16 × 630 = 2,623.2	2,623.2 × 0.095 = 249.20
Annually	49.9	49.9 × 630 = 31,478.9	31,478.9 × 0.095 = 2,990.50

LED tube lighting that can be retrofitted into existing fixtures. All of the 4-foot (1.2-meter) fluorescent lights can be replaced by LED lighting. The usage was estimated for each fixture, and the following table (Table 12.16) shows the energy consumption and total cost of the lighting. Costs were estimated at $0.095 kilowatt hour.

It should be noted that the costs mentioned in Table 12.16 only factor in the purchase price. There are more savings to be made with installation cost reduction and disposal costs, since LED lights do not need to be recy-cled like fluorescent lights.

Cost-Saving Opportunities for Energy Consumption in Entryway Area: The entryway does not have sizable potential, as it is already fairly efficient. The televisions are liquid crystal display (LCD) and use less relative energy as long as they are powered off during the night. Unplugging the TVs would provide further savings, as there would be no standby power being used. Table 12.17 shows the cost of replacing the four footlights with LED lighting and the incandescent lights with typical compact fluorescent light-bulbs (CFLs).[8, 9]

Moreover, the usage was estimated for each of the fixtures, and Table 12.18 shows the energy consumption and total cost of the lighting. Costs were estimated at $0.095 per kilowatt hour.

Table 12.17. Evaluation of Energy Consumption per Time Period (Hour, Day, Month, and Year) and Total Cost of Energy Consumption Related to Entryway Light Replacement

Time period	Energy consumption (kWh)				Cost of energy consumption in $
	Per unit of each LED fixture	Area with LED fixture	Per unit of each CFL fixture	Area with CFL fixture	
Hourly	0.017	0.017 × 14 = 0.24	0.032	0.032 × 4 = 0.13	(0.13 + 0.24) × 0.095 = $0.035
Daily	0.20	0.20 × 14 = 2.86	0.38	0.38 × 4 = 1.52	(1.52 + 2.86) × 0.095 = 0.42
Weekly	1.02	1.02 × 14 = 14.28	1.92	1.92 × 4 = 7.68	(7.68 + 14.28) × 0.095 = 2.09
Monthly	4.25	4.25 × 14 = 59.50	8.00	8.00 × 4 = 32.00	(32.00 + 59.50) × 0.095 = 8.69
Annually	51.00	51.00 × 14 = 714.00	96.00	96.00 × 4 = 384.00	(384.00 + 714) × 0.095 = 104.31

Note that some values are rounded.

Table 12.18. Cost of Replacing Four Footlights with LED Lighting and the Incandescent Lights with CFL Bulbs Related to Entryway Area

Lighting fixture type	Number of bulbs (quantity)	Total wattage in kW = (wattage × quantity)/1000
4-ft LED T12/T8,17W	14	(14 × 17)/1000 = 0.238 kW
135 Watt-equivalent CFL, 32W	4	(32 × 4)/1000 = 0.128 kW

Cost-Saving Opportunities for Energy Consumption in Kitchenette Area: There are a few things that can be done to improve the energy usage in the kitchenette. One such thing is to install a motion-sensor that activates the vending machines. The machines both emit lights, and one provides cool air. The sensor can be installed and the vending machine set to "standby" mode and run on lower power until the switch is activated.

Additionally, the lights that light up the front display may be able to be switched out with more energy-efficient ones; however, our team did not analyze these bulbs. If the machine feels excessively warm to the touch, it may be worth looking into LED lighting. The reason is twofold: The direct energy savings in terms of wattage, but also there would be no excess heat needed to be offset by the cooling mechanism. Small changes such as unplugging appliances at night and during weekends and leaving only one microwave operational can save additional energy. Table 12.19 shows the

Table 12.19. Evaluation of Energy Consumption per Time Period (Day, Month, and Year) and Total Cost of Energy Consumption Related to Reduction in Energy Usage of Kitchenette Appliances

Appliance	Wattage of powered-on mode	Wattage standby mode (estimation)	Energy consumption per time period per unit of each appliance in kWh		Cost of energy consumption related to each appliance located in kitchenette per time period in $	
Dishwasher	1,200	N/A	Daily	2.41	2.41 × 0.095 = 0.23	
			Monthly	20.3	20.3 × 0.095 = 1.93	
			Annually	244	244 × 0.095 = 23.18	
Ice maker	840	N/A	Daily	20.2	20.2 × 0.095 = 1.92	
			Monthly	611.5	611.5 × 0.095 = 58.09	
			Annually	7,338.2	7338.2 × 0.095 = 697.129	
Sharp microwave	1,100	3	Daily	2.2	2.2 × 0.095 = 0.21	
			Monthly	45.8	45.8 × 0.095 = 4.35	
			Annually	550.0	550.0 × 0.095 = 52.25	
Refrigerator	1,800*	N/A	Daily	1.5	1.5 × 0.095 = 0.14	
			Monthly	45.8	45.8 × 0.095 = 4.35	
			Annually	550	550 × 0.095 = 52.25	
Vending machine	200	N/A	Daily	4.8	4.8 × 0.095 = 0.45	
			Monthly	145.6	145.6 × 0.095 = 13.83	
			Annually	1,747.2	1747.2 × 0.095 = 165.98	
Vending machine—(cold + sensor)	200	N/A	Daily	4.8	4.8 × 0.095 = 0.45	
			Monthly	145.6	145.6 × 0.095 = 13.83	
			Annually	1,747.2	1747.2 × 0.095 = 165.98	
Coffee machine	1,800	250	Daily	12.2	12.2 × 0.095 = 1.16	
			Monthly	301.7	301.7 × 0.095 = 28.66	
			Annually	3,620	3620 × 0.095 = 343.9	
Total cost of energy consumption related to all appliances located in kitchenette area per time period in $					Daily	4.11
					Monthly	125.04
					Annually	1,500.67
Total energy consumption related to all appliances located in kitchenette area per time period in kWh					Daily	48.11
					Monthly	1,316.3
					Annually	15,796.6

***The refrigerator usage was estimated, with online estimates of an average annual usage at 550 kWh. The wattage was not used in any calculations.**

potential costs of all these improvements. Costs were estimated at $0.095 per kilowatt hour.

Cost-Saving Opportunities for Energy Consumption in Manufacturing Area: Considering a similar approach to the entryway area, the lights can be replaced with compatible LEDs. Table 12.20 shows the cost of replacing the *U*-shaped fluorescent lighting with 8-foot (2.4-meter) LED, 36W lighting and the 8-foot fluorescent lights with typical 4-foot LED T12/T8, 17W.[10]

Table 12.20. Cost of Replacing Four Footlights with LED Lighting and the Incandescent Lights with CFL Bulbs Related to Manufacturing Area

Lighting fixture type	Number of bulbs (quantity)	Total wattage in kW = (wattage × quantity)/1000
8-ft LED, 36W	224	(36 × 224)/1,000 = 8.064 kW
4-ft LED T12/T8, 17W	214	(17 × 214)/1,000 = 3.638 kW

Table 12.21. Evaluation of Energy Consumption per Time Period (Hour, Day, Week, Month, and Year) and Total Cost of Energy Consumption Related to Manufacturing Area Light Replacement

Time period	Energy consumption (kWh)				Cost of energy consumption in $
	Per unit of each fixture 8-ft LED, 36W lighting	All units of fixture 8-ft LED, 36W lighting	Per unit of each fixture 8-ft LED, 36W lighting	All units of fixture 4-ft LED T12/T8,17W	
Hourly	0.035	0.035 × 224 = 8.06	0.017	0.017 × 214 = 3.64	(3.64 + 8.06) × 0.095 = 1.112
Daily	0.86	0.86 × 224 = 193.54	0.41	0.41 × 214 = 87.31	(87.31 + 193.54) × 0.095 = 26.68
Weekly	4.32	4.32 × 224 = 967.68	2.04	2.04 × 214 = 436.56	(436.56 + 967.68) × 0.095 = 133.40
Monthly	18.00	18.00 × 224 = 4,032.00	8.50	8.50 × 214 = 1,819.00	(1,819.00 + 4,032) × 0.095 = 555.85
Annually	216.00	216.00 × 224 =48,384.00	102.00	102.00 × 214 = 21,828.00	(21,828.00 + 48,384.00) × 0.095 = 6,670.14

The usage was also estimated for each fixture, and Table 12.21 shows the energy consumption and total cost of the lighting. Costs were estimated at $0.095 per kilowatt hour.

Cost-Saving Opportunities for Energy Consumption in Other Areas Containing Restrooms, Hallways, and Conference Rooms: Considering a similar approach to that of the entryway area, the lights can be replaced with compatible LEDs and CFLs. Table 12.22 shows the cost of replacing incandescent lights (39 lamps) with 135 watt-equivalent CFL 32W lighting and the *U*-shaped fluorescent light 4-ft and fluorescent lights (48 bulbs) with typical 4-ft LED T12/T8, 17W lighting.

The usage was also estimated for each fixture, and Table 12.23 shows the energy consumption and total cost of the lighting. Costs were estimated at $0.095 per kilowatt hour.

Table 12.22. Cost of Replacing Current Lighting with LED and CFL Lighting Related to Other Areas Containing Restrooms, Hallways, and Conference Rooms

Lighting fixture type	Number of bulbs (quantity)	Total wattage in kW = (wattage × quantity)/1000
135 watt-equivalent CFL, 32W	39	(32 × 39)/1,000 = 0.918 kW
4-ft LED T12/T8, 17W	48	(17 × 48)/1,000 = 0.816 kW

Table 12.23. Evaluation of Energy Consumption per Time Period (Hour, Day, Week, Year, and Month) and Total Cost of Energy Consumption Related to the Other Area Light Replacement (Such as Restrooms, Hallways, and Conference Rooms)

Time period	Energy consumption (kWh)				Cost of energy consumption in $
	Per unit of each fixture 4-ft LED T12/T8, 17W lighting	All units of fixture 4-ft LED T12/T8, 17W lighting	Per unit of each fixture 135 watt-equivalent CFL, 32W lighting	All units of fixture 135 watt-equivalent CFL, 32W	
Hourly	0.017	0.017 × 48 = 0.816	0.032	0.032 × 39 = 1.248	((0.816 + 1.248) × 0.095) = 0.196
Daily	0.204	0.204 × 48 = 9.79	0.173	0.173 × 39 = 6.78	((9.79 + 6.78) × 0.095) = 1.57
Weekly	1.02	1.02 × 48 = 48.96	0.869	0.869 × 39 = 33.92	((48.96 + 33.92) × 0.095) = 7.87
Monthly	4.25	4.25 × 48 = 204	3.62	3.62 × 39 = 141.33	((204 + 141.33) × 0.095) = 32.81
Annually	51.0	51.0 × 48 = 2,448	43.48	43.48 × 39 = 1,696	((2,448 + 1,696) × 0.095) = 393.68

12.7. Summary of Cost Savings Related to Energy Assessment

Implementing all of the recommendations for the offices, the entryway, the manufacturing area, the kitchenette, and the other areas containing restrooms, hallways, and conference rooms will yield massive reductions in energy. In terms of cost savings per year, excluding the initial upgrade costs, routine maintenance costs, and costs of implementation, this company is looking at a cost savings of approximately $11,560 per year. A breakdown of this approximated cost savings is displayed in Table 12.24. A graph illustrating the reduced yearly power consumption is shown in Figure 12.3.

Because LEDs are a new technology, they are, in turn, more expensive. Costs are likely to go down in time as the demand increases. The price for LEDs represents estimated quantities of purchases of 100 or more. The upfront cost is greater than the benefit in the first year; therefore, the sav-

Table12.24. Summary of Cost Savings Related to Energy Assessment

Summary of estimated energy consumption and costs	Offices	Entryway	Kitchenette	Manufacturing	Other areas
Power consumption (current)	75,124 kWh	4,260 kWh	18,191 kWh	140,832 kWh	15,081 kWh
Power consumption (after replacement)	32,648 kWh	2,058 kWh	15,796 kWh	74,748 kWh	6,568 kWh
Cost to operate (current)	$7,136.78	$404.70	$1,728.22	$13,379.04	$1,432.74
Cost to operate (after replacement)	$3,104.50	$195.51	$1,500.67	$7,101.06	$624.01
Annual energy cost savings	$4,032.28	$209.19	$227.55	$6,277.98	$808.73
Replacement cost (hard costs only)	$35,487.44	$811.36	$200.00	$37,795.36	$3,270.96
Total cost/cost savings considering the replacement costs for the year that replacing has occurred*	$31,455 cost	$602.17 cost	$27.55 cost saving	$31,517 cost	$2,462 cost

***Note that after some years, cost savings will compensate for the cost of replacement. Moreover, numbers and values are based on the conducted analysis in previous sections, adding some other estimations, and original data provided by the company.**

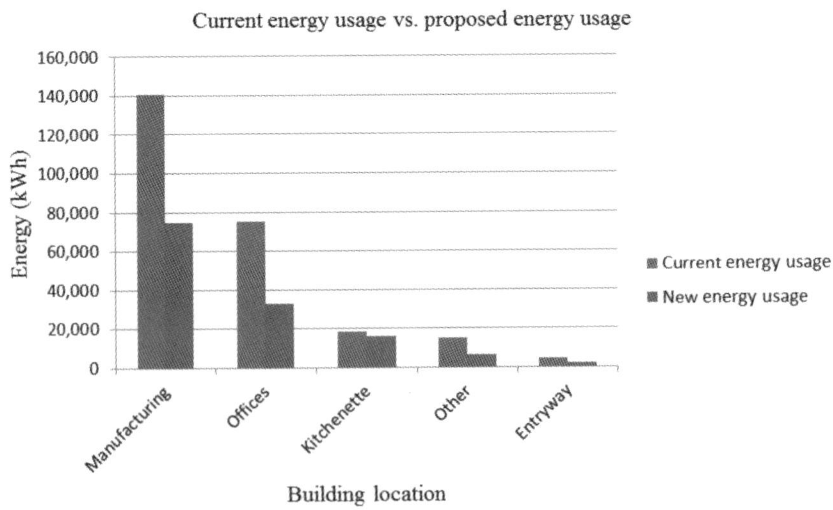

Figure 12.3. Comparing Current Energy Usage with Proposed Energy Usage after Conducting Recommended Improvement Procedures

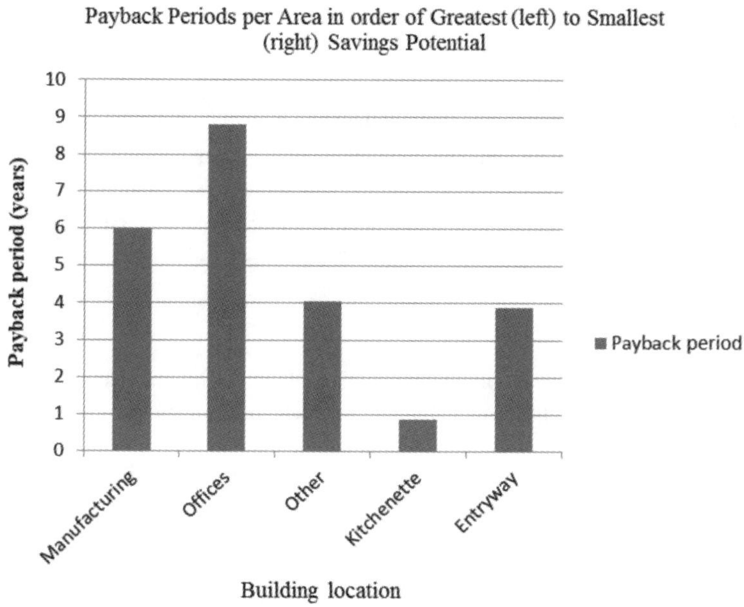

Figure 12.4. Payback Periods Regarding Energy Assessment per Area

ings can be misrepresented. The following graph, shown in Figure 12.4, demonstrates the payback periods within departments.

Based on the results in Figure 12.4, the kitchenette offers the quickest payback period but one of the smallest future savings on energy in accordance with Figure 12.3 and Table 12.24. The manufacturing area has the most potential, as it has a shorter payback period than the offices but also has the most savings on energy out of any area (Figure 12.4, Figure 12.3, and Table 12.24, successively). This could be the effect of the lights being turned on for 24 hours a day, seven days a week.

12.8. Conclusions

Significant environmental and economic benefits were derived from this case study for the company. A total of 7.8 metric tons (8.6 tons) of solid waste per year was identified for diversion from landfill via reduction, reuse, or recycling. A corresponding annualized economic value of $3,205 was also identified through reduced waste disposal costs and revenue from the sale of recyclable materials.

Moreover, based on Table 12.24 great energy reduction opportunities exist for this company. Existing kilowatt hours used per year is equal to 253,488, and by exploiting the full amount of energy reduction opportunities identified, the number of kilowatt hours used per year could be reduced by 131,818. This could result in annual cost savings of approximately $11,560.

References

1. "Toxics Release Inventory (TRI) Program," U.S. Environmental Protection Agency, www2.epa.gov/toxics-release-inventory-tri-program/tri-data-and-tools, retrieved October 20, 2014.

2. Eriksson, O., Finnveden, G., Ekvall, T., and Borkjolund, A. (2007), "Life cycle assessment of fuels for district heating: A comparison of waste incineration, biomass- and natural gas combustion," Energy Policy 35, no. 2, 1346–1362.

3. Bovea, M. D., Ibouez-Forrs, V., Gallardo, A., and Colomer-Mendoza, F. J. (2010), "Environmental assessment of alternative municipal solid waste management strategies: A Spanish case study," Waste Management 30, 2383–2395.

4. Franchetti, M., Solid Waste Analysis and Minimization: A Systems Approach (New York: McGraw-Hill, 2009), 127–137.

5. Government Review of Waste Policy in England in 2011, Department for Environment, Food, and Rural Affairs (DEFRA), www.gov.uk/government/uploads/system/uploads/attachment_data/file/69401/pb13540-waste-policy-review110614.pdf, retrieved October 22, 2014.

6. Evangelisti, S., Lettieri, P., Borello, D., and Clift, R. (2014), "Life cycle assessment of energy from waste via anaerobic digestion: A UK case study," Waste Management, 226–237.

7. "Ranges of Standby Power," Lawrence Berkeley National Laboratory, http://standby.lbl.gov/summary-chart.html, retrieved October 21, 2014.

8. "Information on LED Lighting Products," LEDLight.com, www.ledlight.com/LED-Information.aspx, retrieved October 21, 2014.

9. "Replacing Traditional Bulbs with CFL Lightbulbs," Shenzhen Aodek Optoelectronic Co., Ltd., http://cfllight.com/, retrieved October 21, 2014.

10. Replacing traditional bulbs with more efficient lightbulbs, Creative Lighting Solutions, www.creativelightings.com, retrieved October 21, 2014.

Chapter 13

Case Study: Ball Bearing Manufacturing

This company case study is of a privately owned company that was founded in the 1800s in California and has since grown to become one of the largest producers of ball bearings in the United States. It presently has four manufacturing centers and three service (sales) centers within the United States. The company manufactures a complete range of bearings and cast materials all of which are made to the highest standards to optimize product performance. The company provides its products to industrial and commercial companies from a range of industries.

13.1. Methodology

This section discusses the methodology utilized to conduct the case study. It provides a discussion of solid waste reduction, energy reduction, and hazardous waste reduction.

13.1.1. Solid and Hazardous Waste

The procedure used to estimate the annual solid waste streams at the company involved a sampling of waste containers in several areas. The team focused on three areas: the machine shop, the warehouse, and the office area. By sampling all containers from each area, a general composition was estimated. The containers were each inspected to determine the total volume of waste, types of materials, and quantity of each material. From this information, estimated volumes of each material were calculated for each waste container. Using standard packing densities, weights of recyclable materials were estimated. Volumes are required to determine the waste hauling cost savings, and weights are required to determine the revenue generated from the sale of recyclables. Using the data collected, annual waste streams were calculated.

13.1.2. Energy

The objectives of this project were to define the energy consumption at the company and to identify economically feasible options for energy reduction. During the site visits an energy assessment was conducted and energy consumption data were obtained from the management team. The energy assessment consisted of a comprehensive examination of current lighting fixtures and appliances from the company in the manufacturing, warehouse, and office areas. Energy minimization options for the major areas of energy consumption were researched, with an emphasis placed on energy reduction and economic feasibility. The remainder of this report contains the data summary in addition to options and recommendations that were determined to be most beneficial to the company and to the environment.

Pollution prevention in an industrial environment necessitates in-plant practices that include process modifications, product reformulations, management practices, recycling, and equipment replacements or modifications. A few transformations in management practices helped to meet goals such as the addition of regular staff meetings, the implementation of a preventive maintenance program, training staff, and inventory control. With environmentally friendly buildings gaining traction, there has been a critical thought of utilizing energy-efficient products in the buildings. It is mostly assumed that green buildings are always energy efficient, which is not always true, as a building can attain its green certification based on its other environmental factors and leaving out the energy-efficiency factor. Energy efficiency is important not only because of its advantages surrounding energy use but also because energy efficiency gives the most economic turnover. Annual savings are very high when a building considers an energy efficiency makeover.

13.2. Solid Waste Reductions

The company utilized approximately 56 waste containers. These containers were primarily located in office areas. On average each container was found to be approximately 65% full just before emptying and produced nearly 374 kilograms (825 pounds) per year. Table 13.1 and Figure 13.1 display the weight of major solid wastes disposed of annually by the company. The waste streams are shown by annual tonnage and by the percentage of total waste. As shown in Table 13.1, the facility currently disposes of approximately 101.1 tons (91.7 metric tons) of solid waste each year. Of that total it was determined that 0.6 ton (0.5 metric ton) is currently being recycled, 41.1 tons (37.3 metric tons) are recyclable but not currently being recycled, and 59.4 tons (53.9 metric tons) are waste that are not recyclable in the common markets.

Table 13.1. Annual Waste Stream Generation and Disposal Data

Component	Tons per year	Percentage of total	Tons recycled	Tons not recycled
Mixed office paper	15.0	64.6	6.3	8.7
Nonrecyclable/ food	4.1	17.9	0.5	3.6
PET (1)	1.1	4.7	0.7	0.4
Other	1.0	4.4	0.0	1.0
Aluminum cans	0.7	2.9	0.6	0.1
Magazines	0.5	2.2	0.0	0.5
Cardboard	0.4	1.9	0.0	0.4
Plastic wrap	0.3	1.2	0.0	0.3
Other plastics	0.1	0.3	0.0	0.1
HDPE (2)	0.0	0.1	0.1	0.0
Newspaper	0.0	0.0	0.0	0.0
Glass	0.0	0.0	0.0	0.0
Total	23.1	100.0	8.1	15.0

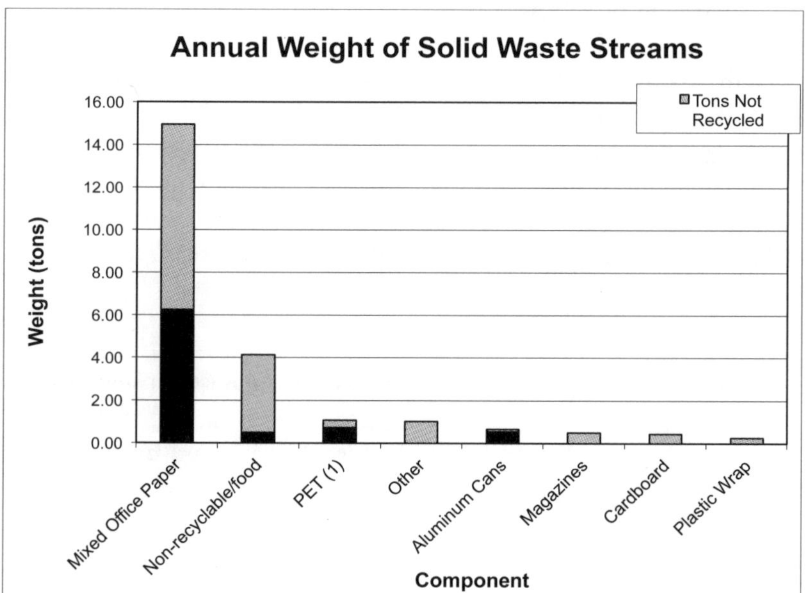

Figure 13.1: Annual Weight of Solid Waste Streams

13.2.1. Waste Hauling Cost Reduction

Table 13.1 and Figure 13.1 outline the major annual waste streams at the company. These charts identify both the recyclable and nonrecyclable elements in each stream. The majority of recycling opportunities for the facility were found in the office areas. Placing the clearly marked recycling containers near existing waste containers in break rooms will help to increase the use of such recycling containers. This will also make it more convenient for janitorial staff to collect the recyclables.

Cost savings could be realized from a reduction in waste hauling costs. Based on the assessments conducted by the waste assessment team, waste hauling costs can be reduced by approximately 45–50% by reducing the volume of waste taken by the hauler and the frequency of solid waste collections. Many waste hauling companies can be contracted to pick up recyclable materials. Most of the volume of waste reduction can come from a more robust paper-recycling program.

13.2.2. Sale of Recyclable Materials

Using the current recyclable commodity prices, the savings displayed in Table 13.2 can be realized from the sale of materials at the company.

It is worth noting, however, that three assumptions were made in Table 13.2. The first assumption is that the material is all baled before sale. The prices will be significantly lower if the material is not baled and many companies will not purchase non-baled materials.

The second assumption is that the prices for recycled commodities will not rise in the coming years. As little as a year ago, mixed office paper was selling for as high as $80 a ton, for example, so the profits from selling recyclables now may be much lower than they will be a year from now, depending on economic conditions.

Table 13.2. Revenue Generated from the Sale of Recyclable Material

Component	Tons per year	Current market value per ton	Revenue (from sale of recyclables)
Mixed office paper	8.7	$10	$87
Newspaper	0	$10	$0
Cardboard	0.4	$40	$16
Plastic bottles	0.4	$10	$4
Aluminum cans	0.1	$400	$40
Total	9.6		$147

Table 13.3. Solid Waste Minimization Recommendations

ITEM	RECOMMENDATION
Improving the Office Recycling Program for paper, plastic, and aluminum cans	Expanding the facility-wide recycling program would eliminate the disposal of tons of solid waste annually. Installing recycling containers in the break rooms and a small paper recycling bin at each desk will make it more convenient for everyone to recycle. Also, placing a recycling container for paper by each printer is a simple way to collect a large amount of mixed office paper. Keeping employees aware of the recycling program is key to the success of the program.
Reducing the frequency of waste hauling	Savings are possible by reducing the frequency of times the solid waste at the facilities is collected. Oftentimes, the waste hauling companies are on a fixed hauling schedule and end up emptying the dumpsters when they are half full, or even less. Especially with a facility-wide recycling initiative, the dumpsters will be less and less full, but waste hauling costs will remain constant. Renegotiation of the contract may be in order.
Using paperless fax machines	Depending on fax usage, substantial paper cost savings could be realized, as well as waste reduction.
Setting printer and copier defaults to print on both sides of the paper, when possible	This will aid in reducing office paper waste. At $5 per ream, the company could achieve significant cost savings as well. An added benefit to this is any reduction in paper use will result in less confidential material going to the document destruction company.
Sending food waste to an energy reclamation facility	This method to reduce the amount of waste going to landfills is not typically a cost effective method due to the shipping costs associated with it. It does, however, remain an option should a reclamation facility to be constructed near the facility's area.
Using biodegradable or recyclable materials for food service	Currently, the use of expanded rigid polystyrene plastic can be reduced by switching to paper plates and cups or using a vegetable oil–based material instead of a petroleum-based one, which will be better for the environment. There are many cost-competitive alternatives to polystyrene plastic.

The third assumption is that all of the potentially recyclable materials are captured. This is not an entirely practical assumption, but the actual amount that will be is extremely hard to determine and depends on the placement of recycling containers, the display of the containers, and employee awareness of the program. The values displayed in Table 13.2 are the goals to aim for but should not be used for planning decisions. Depending on the current value of the recycling materials on the commodity market, it may or may not be cost justifiable to try to gather the recyclables in order to sell them for a profit. Typically, values per ton fluctuate between $20 per ton to $100 per ton on the commodity markets.

13.2.3. Recommendations for Solid Waste Minimization

Table 13.3 provides a list of recommendations for the company to minimize solid waste disposal. The items and a detailed description of each are provided.

13.3. Energy and Hazardous Waste Reductions

The company utilized 583 high energy–consuming bulbs, which were analyzed for the assessment. These bulbs can be replaced with alternative lighting fixtures without compromising on the lighting output while consuming lower energy. Tables 13.4 and 13.5 provide details regarding the current lighting energy consumption and cost per year to operate. For the calculations, the cost of energy was $0.10 per kilowatt hour.

Table 13.6 provides alternative lighting fixture data, Table 13.7 provides the total cost per year for each recommended lighting fixture, and Table 13.8 provides the approximate savings the facility can obtain by changing lights.

Table 13.4. Lighting Fixtures Data from the Facility Used in Energy Assessment

Fixture type	Fixture details	Wattage	Quantity	Operating hours per year
A	Incandescent lights (Machine Shop)	400	155	2,125
B	High pressure sodium bulbs (Warehouse)	400	66	2,125
C	Fluorescent bulbs T12	40	168	2,750
D	Fluorescent bulbs T8	34	101	2,750
E	Fluorescent bulbs T8	32	93	2,750

Table 13.8 gives the information about the cost savings by replacing existing lighting fixtures with more energy-efficient fixtures.

Table 13.9 provides the information about the amount of greenhouse gases generated, the energy used, and the hazardous waste produced with the current fixtures and better alternative fixtures.

Table 13.5. Total Cost per Year for Each Existing Lighting Fixture

Fixture type	A	B	C	D	E
Wattage	400	400	40	34	32
Quantity	155	66	168	101	93
Total wattage	62,000	26,400	6,720	3,434	2,976
Total hours/year	2,125	2,125	2,750	2,750	2,750
kWh/year	131,750	56,100	18,480	9,444	8,184
Cost/year*	12,516.25	5,329.50	1,755.60	897.13	777.48

Total kW/year =	102
Total kWh/year =	223,958
Total cost/year = $	21,275.96

Assumptions:

1. ***Number of hours per day machine shop and warehouse are operating: 8.5***
2. ***Number of hours per day office areas are operating: 11***
3. ***Number of days per year facility is operating: 2500***

Table 13.6. Alternative Lighting Fixture Data

Fixture type	Fixture details	Wattage	Quantity	Operating hours per year
A	Compact fluorescent lights (CFL) (Machine Shop)	80	155	2,125
B	Compact fluorescent lights (CFL) (Warehouse)	80	66	2,125
C	Fluorescent bulbs T5 (Office)	28	168	2,750
D	Fluorescent bulbs T6 (Office)	28	101	2,750
E	Fluorescent bulbs T7 (Office)	28	93	2,750

Table 13.7. Total Cost per Year for Each Recommended Lighting Fixture

Fixture type	A	B	C	D	E
Wattage	80	80	28	28	28
Quantity	155	66	168	101	93
Total wattage	12,400	5,280	4,704	2,828	2,604
Total hours/year	2,125	2,125	2,750	2,750	2,750
kWh/year	26,350	11,220	12,936	7,777	7161
Cost/year = $	2,503.25	1,065.90	1,228.92	738.82	680.30

Total kW/year =	28
Total kWh/year =	65,444
Total cost/year = $	6,217.18

Table 13.8. Approximate Savings the Facility Can Obtain

Fixture type	Current fixtures' annual cost ($USD)	Savings opportunity	Recommended fixtures' annual cost ($USD)	% reduction in cost	Annual Savings ($USD)
A	12,516.25	Reduced wattage	2,503.25	80.00	10,013.00
B	5,329.5	Reduced wattage	1,065.9	80.00	4,263.60
C	1,755.6	Reduced wattage	1,228.92	30.00	526.68
D	897.13	Reduced wattage	738.82	17.65	158.31
E	777.48	Reduced wattage	680.3	12.50	97.18
Total	21,275.96		6,217.19	70.77	15,058.77

These reductions observed in the Economic Input-Output Life Cycle Assessment (EIO-LCA) results are attributable to actual reduction in electricity usage, not to replacements of bulbs. Tables 13.10 and 13.11 are the outputs of a greenhouse gas (GHG) conversion tool and a pollution prevention (P2) cost calculator, respectively.

Table 13.9. Results of Environmental Input-Output Life Cycle Analysis Output for Cost Savings Associated with Energy Savings per Year

Factor	Savings ($USD)
Greenhouse gases (metric tons of CO_2 equivalents)	141
Energy (terajoules)	1.68
Hazardous waste generation (short tons)	1,890

Table 13.10. Pollution Prevention Greenhouse Gas Conversion Tool Output

State	Electricity conserved (input value)	Unit reported	Electricity conserved (kWh)	GHG reduction ($MTCO_2e$)
California	158,514	kWh	158,514.00	141.24

Table 13.11. Pollution Prevention Cost Calculator Output

Electricity conserved	Unit (select)	Unit cost ($/unit just selected)	kWh reduced	Dollar savings
158,514	kWh	0.0950	158,514.00	**15,138.09**

13.3.1. Energy and Hazardous Waste Results

The tables discussed in this case study indicate that savings can be obtained from switching the current lighting fixtures to energy-efficient fixtures. Based on the assessments done using the Energy Assessment Spreadsheet (EAS) tool created by the research team, the cost savings obtained were $15,058.77 while reducing electricity usage from 223,958 to 65,444 kilowatt hours. This was 70% of the original energy bills generated by the present lighting fixtures. These savings have been obtained just by replacing the existing 400W incandescent and high-pressure sodium bulbs in the machine shop and the warehouse 80W compact fluorescent bulbs and also by replacing the T12 and T8 bulbs in the office areas with T5 bulbs. It can be concluded from the EIO-LCA output that the emission of greenhouse gas reduction of 141 $MTCO_2e$ can be achieved by changing lighting to more efficient options. Also it was observed that there are nearly 70% less hazardous waste emissions when energy-efficient lighting fixtures were adopted. The output from the GHG conversion tool suggests that there is a reduction of 141.24 $MTCO_2e$, while the P2 cost calculator gives the cost savings as $15,138.09, which is in accordance with the output obtained from the EAS tool.

13.4. Conclusions

Significant environmental and economic benefits were derived for the company from this case study. Specifically, the following annual reductions were achieved: 13.6 metric tons (15 tons) of solid waste, 158,000 kilowatts of energy, 141 metric tons (155 tons) of greenhouse gas emissions, 1,714.2 metric tons (1,890 tons) of hazardous waste, and over $15,000 in cost savings.

Chapter 14

Conclusions and Final Insights

Sustainability and environmental protection have emerged as core business requirements for manufacturing companies from a market share, legal, and customer expectation point of view. Legal systems and environmental regulations are strengthening and requiring manufacturing companies to place a stronger focus on minimizing environmental footprints. To assist in the process, many models, strategies, and tactics have emerged for manufacturing companies. As discussed in this book, these range from forecasting models to life cycle analyses and risk assessments to optimization models.

Manufacturers need to place emphasis on all components of their environmental footprints, including air, water, and solid waste. These aspects will not only better protect the environment but also generate positive economic value, as discussed, regarding recycling commodity markets, energy reduction, and alternative energy implementation. From a business standpoint, traditional activities such as inventory control and transportation optimization offer both environmental protection in terms of fuel/emission reductions and significant cost reductions.

The need for sustainable operations will only increase and intensify in the future. World population is still growing at about 80 million people a year, or about 220,000 people per day; if current trends persist, there will 2.5 billion more people on the planet by mid-century, bringing the total to about 9.2 billion.[1] Providing clean water, ample food, and other resources for this expanding population from resources on a finite planet pose a major challenge. Through sustainable operations and conservation, these concerns can be addressed.

As we progress into the future of manufacturing processes and systems, these issues will intensify and a focus on sustainability will increase. Below is a list of expected future trends in these areas:

- More legal regulation for developed and developing countries related to emissions, hazardous waste, and solid waste generation.
- The integration of global sustainability initiatives for international corporations; companies will apply equivalent environmental policies in all countries of operation based on the most stringent regulations.
- High competition from globalization and the need to drive down costs.
- High consumer focus on sustainable companies.
- Higher visibility of errors owing to social media and the rapid dissemination of information.

Sustainability and environmental protection are no longer just an option for manufacturers; they are now expectations for doing business.

Reference

1. "Why Population Matters," The Population Institute, www.populationinstitute.org/resources/whypopulationmatters/, retrieved September 16, 2014.

Index

Note: *f* in page numbers indicates a figure; *t* indicates a table

F

J

K

L

M

S

T

U

V

W

X

Z

About the Authors

Matthew Franchetti is an Associate Professor of Mechanical, Industrial, and Manufacturing Engineering and the Director of Undergraduate Studies for the Mechanical Engineering Program at the University of Toledo in Ohio. He is the Director of the Environmentally Conscious Design and Manufacturing Laboratory and the Principal Investigator for the Business Waste Reduction Assistance Program, a joint effort with local government. Prior to his work in academia, Dr. Franchetti served as an industrial engineer and technical manager for the U.S. Postal Service. A Certified Six Sigma Black Belt and a professionally licensed engineer, he has helped numerous companies reap the benefits of sustainable practices.

At the time of this publication, **Behin Elahi** is a Ph.D. candidate in Industrial Engineering at the University of Toledo, having received her M.S. in this field in 2011. Her research interests include green supply chain, operation research, quality control, and mathematical programming, and she has published more than sixteen conference and journal papers. Since 2012, Elahi has been a consultant and one of the project managers of the Business Waste Reduction Assistance Program under the supervision of Dr. Franchetti. She is an executive officer for the Institute of Electrical and Electronics Engineers (IEEE) Toledo section and a member of the University of Toledo Engineering Council (UTEC), and the Society of Women Engineers.

Somik Ghose also is pursuing his Ph.D. in Industrial Engineering at the University of Toledo. He completed his M.S. in the field of Civil Engineering in 2003 and has worked for eight years as an engineer and manager at the Center for Innovative Food Technology/Edison Industrial Systems Center, one of Ohio's Manufacturing Excellence Partnership (MEP) centers. Ghose has participated in projects involving implementation of water and wastewater improvements, energy efficiency, and renewable energy technologies in the automotive, manufacturing, and food-processing industries. He also has been involved in collaborative research and development regarding biofuels and waste-to-energy processes.